DESIGNING BRIDGES TO BURN

DESIGNING BRIDGES TO BURN

Architectural Memoirs by Stanley Tigerman

ORO editions

CONTENTS

Many dedicate their autobiography to their significant other. In my case there is a particular rationale for me to do just that: Margaret McCurry edited the text. While I think I deserve a medal for having suffered through a spouse's ruthless alterations, it is small potatoes next to the generosity of her volunteering to do it in the first place.

FOREWORD
The two characters of an autobiography

BY EMMANUEL PETIT

No document of Stanley Tigerman can represent his persona and his views on architecture more authentically than an autobiography, inasmuch as the architecture of this sort of text is constructed from the vantage point of the first-person narrative. Without a doubt, one of Tigerman's most idiosyncratic and cardinal contributions to the architectural discussion has been his relentless insistence on the centrality of the ethical character of the architect in interpreting his or her physical, cultural and sociological habitat; as such, every single one of Tigerman's utterances bears the traces of the author's individuality, and is, in a way, to be seen as a transcript of his personal life story. He has unconsciously aligned himself with the claim made by humorist and satirical novelist Evelyn Waugh that autobiography "is not the most important subject in history but it is one about which you are uniquely qualified to speak." This being the case, the autobiography allows the reader to partially circumvent the official libretto of "Tigerman" in order to also glimpse into the more picaresque scrapbook of "Stanley"; without this insight into Stanley's weltanschauung, the architectural thematic of Tigerman will remain inscrutable.

The "heroes" of modern architecture, like Ludwig Mies or Charles-Édouard Jeanneret-Gris, had reinvented their proper names to be able to speak about themselves in the third person; had not Jeanneret announced his ambitious vision to be reborn as a Nietzschean overman already at the age of 22 when he sent a Christmas card to his parents with a self-portrait as a vulture—the allegory of "spirit"—alongside the message: "The misery of living makes man! And the disdain of this misery of living is incarnated in the soul of the grand Condor." From this moment on, Jeanneret turned into the quasi-prophet Le Corbusier and related of his philosophy in the third-person narrative voice, as he did in his 1932 book Crusade, Or, The Dawn of the Academies: "Leaning against the railing, Le Corbusier thought: One must search the light. Will there be a light source to shine on those stormy roads of this new architecture, which will bring about the new century?"

Stanley has also used an invented name, the pen name Morris Lesser, but unlike Mies and Jeanneret, he did it with much the same intentions with which Søren Kierkegaard, the father of existentialism, alternatively wrote as the pseudonymous characters Victor Eremita, Johannes de Silentio, Johannes the Seducer and Anti-Climacus, among others; for Stanley, the story of architecture was fundamentally multifarious, polyphonous and non-reducible to grand "catholic" styles and tenets. And so, just like

Kierkegaard made of the idealist Systemdenker G.W.F. Hegel his intellectual other, Stanley saw in the German emigré Mies van der Rohe his Oedipal (modern) alter ego. More voodoo doll than straw figure, however, Tigerman's version of Mies was contrived to release Stanley from a spell, which he felt was cast on his generation of Chicago architects by the rigid universals and the zeitgeist ethic of (post-"Hellenic") modernism. This father figure could not be killed, but had been set adrift in the endless ocean of the architectural subconscious of Stanley's generation of (post)-modern Chicago architects—the Titanic, alias Mies, could not simply sink, but instead had to stay suspended on the ocean surface as the paragon of a modern belief system, whose framework has irretrievably been corroded.

Anglo-American poet W.H. Auden once alleged that, "every autobiography is concerned with two characters, a Don Quixote, the Ego, and a Sancho Panza, the Self." Stanley Tigerman truly embodies Auden's incongruous and complementary double character—which is not only symptomatic of postmodernism's complicated relationship towards the past (and modernism in particular), but is also suggestive of the essential divide inhabiting the "ethical" person, that is, the gap between the aesthetic and ethical stages of existence, which Kierkegaard had marked out in his dual tome Either/Or. Is not the myth of architecture's "knight of purity," Mies van der Rohe, Tigerman's autobiographical Don Quixote, and "Stanley" his Sancho Panza? One can discern, in Tigerman, Cervantes's dyad mentalities: On the one hand, Tigerman inherited from both Mies and from Paul Rudolph the architectural instinct to join the "great adventure" of a modern formal language (and become, in Rem Koolhaas's terminology, "a voluntary prisoner" of the discipline); yet on the other hand, one observes Stanley's relentless itch to overturn Mies's delusional enchantment with a particular brand of skepticism, alternatively boosted by the satirical comedy of Charlie Chaplin, the surrealist wit of René Magritte, the jaunty irony of Claes Oldenburg or the libidinal analyses of Sigmund Freud, among others. The Janus-faced divide put forward by Auden does never go away with Tigerman, but paradoxically constitutes the connective motif throughout his work—from his writing and his painting, to his travel sketches and cartoons (Architoons), to his furniture, jewelry and tableware designs, and to his projects on the architectural and urban scales.

Two beliefs are vital to Stanley's world view: On the one hand, he insists on the "Americanness" of his work—a conjecture, which is never divorced from a commitment to the ideas of individual freedom and to a certain liberalism of thought and taste; Vietnam, pop art, suburbia, cowboys and Indians are only some of the American archetypes that Stanley won't relegate to a cultural sphere "outside" of architecture. On the other hand, to see architecture as "a failed attempt at healing an irreparable wound," as he always has, denotes the religious and Cabalist dimension of his work; as a Jew in Chicago, Stanley has cultivated the existential and erratic "drift" between the contrasting sentiments of alienation and belonging, which Saul Bellow so captivatingly described in his The Adventures of Augie March; the bildungsroman begins with the famous sentence: "I am an American, Chicago born—Chicago, that somber city—and go at things as I have taught myself, free-style, and will make the record in my own way: first to knock, first admitted;

sometimes an innocent knock, sometimes a not so innocent. But a man's character is his fate, says Heraclitus." In Tigerman's spiritual perspective, the rational principles of the Jesuit Laugier's primitive hut, concretized in the extant temple of Hellenic culture (the Parthenon), give way to the collective, Jewish memory of the vanished Temple of Solomon. The unfailing awareness of the imperfection and finitude of all human "architecture" haunts Tigerman's work, and the inversion of the conceptual focus from presence to absence (or loss), remains one of its central leitmotifs.

Tigerman is an outsider to architectural reason in the way Socrates or Kierkegaard are outsiders to philosophical logic. Here is the life story of a person who understands the limitations of solutions and answers, and instead likes to draw attention to the complexities of questioning. Architecture does not heal every problem, but makes itself the long-lasting index of a finite human presence. His regard for unresolved dialectics nurtures an ethics of alterity, which eschews the "control" of the subject/architect; did not Talmudic scholar and philosopher Emmanuel Levinas maintain that, "if one could possess, grasp, and know the other, it would not be other." Tigerman's theory of architecture is, therefore, never detached from the autobiographical quest of an individual who tries to situate himself in a world he has been thrown into. Nevertheless, he suggests in this writing that a person's place is not defined by the material boundaries of the buildings and cities he inhabits, but by a different type of "tectonics": the (ethical) encounter with the "Other."

New Haven, March 2010

INTRODUCTION

It is an understatement to propose that I have had misgivings about undertaking these memoirs. Not only have I not had an exemplary life, but I have always felt that I was an outsider—an auslander—to boot. After all, where would one find enough interested readers to justify a book-length dissertation about an ill-advised life in a field—architecture—that, perhaps more than any of its other characteristics is one that demands a normative trajectory to be on an inside track? On the other hand, I have had more skewed trajectories and misguided tangents than I care to remember.

Any career path has its share of hiccups and missteps, but if the time I've spent speaking without thinking, correcting mistakes I made willfully (perhaps to retain my outsider pose), or getting back on track from either premature or misdirected career derailments had been put into distilling my work, I can't imagine how I might have embellished my life as an architect. Who knows, perhaps not much would have changed no matter what direction my career might have taken or how goal-directed my ambitions found their voice.

As it is, it took me three years of uncharacteristic ambivalence to embark on this project. If there is a reason that I eventually decided to begin scratching away at my own skin, it was based on my assumption that while I'm not anyone's idea of a traditional role model, my kaleidoscopic veneer is made up of some unusual facets that might plausibly be scrutinized by those considering a career in architecture irrespective of their trepidations. When I came to the realization that there might be value in compiling a bildungsroman—a kind of memoir with attitude—I was hooked. Who can say if some readers might prefer to play while others already in the game might opt out?

Over time, my own idiosyncratic search for what I would refer to as ineffability (what the late 20th-century architect/educator John Hejduk referred to as aura and what others might simply refer to as indescribable) may bear on those whose interest in architecture could be understood as something more than a useful, even geometric or aesthetic, art. Indescribability taken to its extreme was once referred to by the architectural historian/theorist Tony Vidler as the unheimlich, that is, unhomely or uncanny—as comfortable a place for an outsider to inhabit.

In all events, the poetic aspect of architecture is what fascinated me initially. It took a lifetime to ascertain if poetry was embedded in space or in mass, but that search has kept me engrossed for over six decades. If my career was kaleidoscopic, so is the following text. Rather than guide the reader along a straightforward chronological path, I have chosen to follow a more complex thematic route to storytelling.

*   *   *

I should state at the outset that architects have a well-deserved reputation of being self-centered and aloof. Like most doctors and many lawyers, certain architects are overwhelmed by their own gibberish, which in turn can be perceived as being elitist. Without excusing such conduct, the very nature of conceiving "plans," that is, looking down from above while creating those pocket-sized drawings that later will be used to erect structures that are far larger than the drawings that initially represented them can begin to explain, if not to justify, why it is that architects are sometimes accused of "playing God."

Another forewarning: If my language sometimes seems to pontificate, that posture comes naturally for an architect. Because it is not in my DNA to accept aesthetic rationalizations as the be-all and end-all of conceptualization, I am among those in the field who feel the need to hold some sort of belief system in order to have a raison d'être for what they design. It is but one short step from having a belief system to holding everyone hostage to it. Thus, if what follows expresses some of the traits to which I am alluding, forgive me, but for some of us it comes with the territory.

*   *   *

Whatever presumptions one might harbor that architecture is approachable only through conventional channels navigated sequentially and validated by tradition (the inside track), this manuscript might suggest that such predilections can be unnecessarily limiting. In any case, such old-fashioned preconceptions are not the only way to approach the world of conceptualizing form, let alone testing those concepts by attempting to fabricate them.

For example, the concern that structural engineering is obfuscated by one's fear of the impenetrability of advanced mathematics courses such as integral calculus and differential equations can be readily unpacked when one is brought to the realization that understanding the inherent nature of materials that define architectural structures is ultimately more intuitive than quantitative. An insight into the nature of gravity is far more important than being able to express one's knowledge of the specifics of materiality and its structural quantification. It is more important that an architect instinctively knows the size of some member made of a particular material that might reasonably span a certain distance than it is to master the mathematical formulae that will more precisely determine such material selections. In other words, while the cadaver might look a bit odd, there is perhaps more than one way to skin a cat.

*   *   *

What follows should encourage those who might not otherwise think that they had "the right stuff" to pursue a career in the design of buildings. Characteristics such as diplomacy, patience and the staying power to focus for a half-century and more on a life-long pursuit as demanding, albeit as stimulating, as architecture and remain enthusiastic throughout are the predictable self-assessments thought indispensable to succeed in design. As this book will attempt to explain, other not-quite-so-acceptable traits might, unpredictably enough, also be useful to the fledgling architect trying out his or her wings in a field that at first might not appear to be suited to their individual, perhaps even idiosyncratic capabilities.

Upon reflection, I believe that my decision to become an architect was the right one for me, irrespective of the countless detours that the process took. In hindsight, however, I don't particularly recommend to anyone wishing to amble through this field the several unrelated roads down which I meandered. Surely, there are more discrete, perhaps even more direct ways toward the end of accomplishment than the ones described here. But this text is more a reporting of what transpired than a recommendation that any one way of approaching a career in architecture is more relevant than any other.

Thus, in a certain sense, this is not a "how-to," but rather a kind of "how-not-to" book. In the decision-making process, I made enough mistakes to entertain even the most casual reader, let alone those fastidious enough who might seriously consider pursuing architecture as a discipline.

*   *   *

Traditionally, a person who is able to maintain a lifelong career in "the-mother-of-the-arts" is usually buttressed by a long attention span combined with a high threshold of pain. My wife and partner, the architect Margaret McCurry, is just such a person even though she also came to the discipline as an outsider of sorts. Doggedly tenacious, Margaret will track down the absolutely appropriate, as well as the outright optimal solution to an architectural problem through hell and high water and then cause the result to come to light simply by exercising sheer will. Her conviction about her own innate sense about the aesthetics of design is such that once focused, she is capable of doing whatever it takes to realize her dream.

While there is a reassuring consonance in her goal-directed architectural career that is embedded in her meticulous approach to problem solving, my own impulsive architectural processes are problematically situated somewhere between scatter-shot and hit-or-miss.

Bruce J. Graham speaking at the Art Institute of Chicago on the occasion of an Architecture and Design Society event roasting the author, Chicago, Illinios, 1990)

My individual interpretation of Margaret's resolute way of contending with, let alone overcoming, resistance to her initiatives is that she doggedly obsesses about problems that remain unresolved. While I understand that to be obsessive is not precisely synonymous with tenacity, my own behavior in such circumstances is the product of a bull-headed personality. Unlike those who might look for a chink in the armor of some immovable object resisting their will, I tend to meet that object head on and attempt to overturn it then and there. My bullheadedness on this subject is utterly opposed to what the cult figure and author Edward de Bono posited about being flexibly disposed when one is otherwise stymied. While de Bono's 1970 book, *Lateral Thinking: Creativity Step by Step*, explains that it is often better to go around the wall that stands in one's way than to assault it frontally. Unfortunately that's just not my way of problem solving.

*   *   *

Architects come to understand early on that "willing buildings into existence" is not for the faint-of-heart. Bruce J. Graham, the alpha male architect and previous Skidmore, Owings & Merrill (SOM) general partner in charge of design in the firm's Chicago office and my former boss in the last 18 months that I was employed at SOM, once literally drilled that fact into me one afternoon as the two of us purposefully marched down the street towards the SOM office. With a forefinger setting up a staccato rhythm upon my sternum, he forcefully averred that "architecture is not for pussycats." By that emphatic statement, I understood that he meant that becoming an architect required a seriousness of purpose, combined with an indomitable resolve to grasp whatever ends come into one's purview and then guide them forcefully, and in whatever way necessary to their inevitable, irrefutable conclusion.

Bruce Graham's determinist view of architecture has obvious merit, particularly for those seeking some sort of resolution through the vehicle of their work. Nonetheless, I was never sure how an architect who might bring delight into people's lives fit into his closed system. Perhaps these dialectical opposites didn't allow for the possibility of being present simultaneously in a firm like SOM whose strident seriousness of purpose seemed to transcend über alles. Certainly, the poetic aspects of architecture never occupied that part of Bruce's mind that he conveyed to his minions at SOM other than to aver that "this" (normally his design solution) was better than "that" (normally the design solution of someone else).

Under most, if not virtually all, circumstances, designing buildings is not for the gadfly; Philip Johnson's idiosyncratic late 20th-century example of flitting from style to style after others had initiated a direction that he subsequently appropriated notwithstanding. The long-term rewards that come to those who diligently pursue one overarching course should be sufficiently satisfying so as to supersede seeking short-term returns that have little value beyond their own immediacy. On the other hand, to qualify for such long-term gratification required a higher degree of staying power than I was ever able to dredge up from within.

Nevertheless, my intractable determination to fulfill my own self-endowed prophesy that I would somehow prevail and become an architect overcame anything and everything that had the audacity to stand in my way, including my own desire for immediate gratification. I didn't understand it at the time, but my obstinacy on the subject would carry me through all of the years ahead through obstacles unbeknownst to me in the early stages of my inflexible understanding of architecture.

*   *   *

Practicing architecture is made more difficult for those whose motivation is first and foremost profit. In a free-market-based capitalist society, there are far more rewarding ways to unearth financial remuneration than by becoming an architect. For example, if ethical behavior is not going to be troublesome, any and all kinds of entrepreneurial business practices come to mind. In any case, the business component of architectural practice tends to diminish the discipline itself, particularly the "ambulance chasing" aspects of marketing and branding, both of which privilege the seeking of work over and against discovering the implicit value that might conceivably be embedded in the work itself.

The amount of energy and time expended combined with the commitment required to build well often doesn't permit the work itself to be especially lucrative, at least in the context of a qualitatively driven practice. Of course, that assumes that building well is your goal and that the amount of time you are willing to spend on design is unlimited.

The art of architecture is not for those who are offended by elitism, since recognition within the field tends to be limited to the crème de la crème. Focusing on the work should be fulfillment enough without the angst connected with acceptance or denial to available reward systems.

In a holistic sense, architectural practice demands that conceptualization requires concretization for ultimate fulfillment, which in turn suggests that a highly developed interest in the science as well as in the art of building be acquired.

*   *   *

The schism between the art and the science of architecture was first codified institutionally in continental Europe; particularly within the Austro-Hungarian Empire and then subsequently in those German educational institutions constituted to prepare architects for professional practice. For example, the Academie von Baukunst was a method of education for those interested in an aesthetically derived approach to architecture, while the Technische Hochshule offered an alternative method that looked to a more engineering-oriented approach for inspiration.

In a similar fashion in the United States, some architecture schools whose predilections favored enculturation, such as Ivy League colleges, proposed a thorough orientation in the liberal arts before matriculating into a graduate professional program. Other schools that were more pragmatically inclined, such as land grant colleges, favored engineering among other usefully derived course work as the precedent of choice for those matriculating into a professional program directly following high school.

In the process of establishing the appropriate pedagogy for these two very different types of architectural education, a dialectic schism evolved that was internally focused on validating each method precluding the possibility of any continuous dialogue between the differing aspects of design pedagogy. Architectural education would have benefited enormously from an intersection between these two methods of training far beyond what was thought to be useful to each.

One could argue that an aesthetically driven approach favored space over mass, while a technically informed approach favored mass over space. One could further argue that the intersection between these two dialectically differentiated pedagogies suffered somewhat between architecture as a fine art versus architecture as a useful art.

*   *   *

Then there is the friendly art of persuasion that is itself something of an art. Unlike other visually inclined fine artists who have successfully operated independently, e.g., painters and sculptors, the labors of those committed to the useful arts such as architecture, textiles, industrial and graphic design are not produced in isolation. Decisions about virtually everything connected with their work are jointly and severally derived and require concurrence, perhaps even argumentation, but in all events participation to bring to fruition and are required by the many over and against the few who have a vested interest in a given project. In other words the design arts generally and architecture specifically benefit greatly from democratically driven team play—not to diminish the work of the individual within those teams.

Alas, architecture can be a frustrating field to those who engage solely in academic and/or theoretical constructs. These academicians help to define the evolution of the discipline, albeit "architecture interruptus" without the necessity, indeed without the obligation for their ideas to be put into practice. Those who build sometimes interpret those who think as being scolds, while those who posit ideas sometimes see those who build as unthinking. This unresolved couplet adds little to an otherwise necessary dialogue.

Like mastering the piano, architecture also requires practice; though perfection, as an end in and of itself, is virtually never achieved. But the poetic value that is expressed in the iconic figure of the Vitruvian Man always hovers over those designers who would give credence to that which is thought to be relative versus that which is considered absolute. Understanding the difference between relative and absolute value as an ethical construct is essential to the designer seeking a raison d'être to his or her architectural production.

In the same spirit that "golf is not a game of perfect," over time, I too have belatedly discovered that architecture is also not a game of perfect. But perfection, while perhaps a not-unreasonable aspiration for those who are compulsive, is not apriori the most rewarding aspect of architecture. My desire for perfection is an indication that I always favored an ideal, over and against those who think that architecture is a matter of relativity.

Like Zen Buddhism, being at one with one's work should satisfy whatever need for gratification one may require by dint of providing an inner peace. Unfortunately, I am not one of those people.

*   *   *

Akin to the prizefighter who spars endlessly before daring to step into the ring, the architect studies the nature of materials through his or her many years of apprenticeship before attaining the courage to subject their designs to the elements.

In 1938, on the occasion of Mies van der Rohe's emigration to the United States to accept the position of director of the Architecture School at the Illinois Institute of Technology, the wily Philip Johnson escorted Mies to Frank Lloyd Wright's summer studio, Taliesin, in Spring Green, Wisconsin so that the two giants could collide. While Philip had been the curator of the architecture department at the Museum of Modern Art, his ambitions to practice the craft that he had helped to bring to the public's eye in his seminal exhibition on the International style were known to all who met him.

Mr. Wright is reported to have sarcastically asked Philip "if he was still considering putting things out in the rain." The great man was no doubt trying to intimidate Philip by alluding to the intensely responsible nature of building, wondering aloud whether Philip Johnson had been interested in becoming an architect long enough to be so bold as to build. Of course, Mr. Wright's challenge was altogether facetious since, as a fledgling architect, he built as soon as he acquired a commission proving that the adage "do as I say, not as I do" was central to Wright's architectural lexicon.

Nonetheless, in the architectural world at large, patience is generally perceived of as a virtue. In the field of architecture, patience and persistence are both thought to be fundamental to the definition of practice. But these aren't the only requirements that define the discipline. Inspiration is at the heart of practice since it is understood as basic to the generation of form. One might think that design is such an inspiration but I would contend that the search for a poetic component is at the heart of an architect's desire to define the essential nature of his or her work within the field.

Thus, patience versus inspiration are only two among many dualistic, often contravening conditions that architects need to embrace simultaneously. Is it any wonder that architects are sometimes accused of schizophrenic behavior? And speaking of schizophrenia, the medical definition of that malady is the ability to rotate objects in the mind; the definition of which architects attribute to themselves on a regular basis.

*   *   *

Architecture is at its most challenging and ultimately at its most satisfying when it is understood as a calling. In that sense as in Umberto Eco's opus *The Name of the Rose*, architects are metaphorically analogous to the monks laboriously transcribing the Bible by hand in the scriptorium in their Cistersian monastery at Melk, Austria, before Gutenberg invented the printing press. Even as they labored trying to replicate their illuminated manuscripts, Eco depicts the monks as inescapably erring in the process of copying the Bible, but their mistakes did not seem to discourage them from pursuing their goal of spreading the word of God. Because of their rock-solid faith, those same monks might argue that it was sufficient to simply read the Bible rather than to be so bold as to interpret it. By extension, one could argue that because of a shared belief in the importance of poetry as it relates to practice, those who study the discipline of architecture might consider learning sufficient without the obligation of practice.

From an architectural point of view, however, the architect's designs can be every bit as illuminating when the process is not only revealed but celebrated, and when the product is the result of careful cogitation it needs to be built artfully and well.

Similarly, as revelations as to how buildings are assembled become apparent to the novice architect, the unveiling of process is akin to understanding from whence matter originated. In the mid 1970s, I proposed teaching a design studio at Yale about the architectural implications of an abattoir. Cesar Pelli, then dean of the Yale School of Architecture, thought it a wonderful idea insofar as his own parents had told him that, "if you consume something, you need to know from whence it came, and how it got to the table."

Cesar's wife, Diana Balmori, thought it "a ghastly idea." My aspiration was that those Yale students who signed up for the studio would be exposed to, and hopefully understand, the metaphoric relationship between the transformation of certain living matter into food and how other living matter is transformed into the materials that make up the craft of building. I had hoped that the primal nature of the premise of such a design studio would be enough to make my point about the importance that matter had on the design of buildings. In the context of the potential that matter might contain the seeds of ineffability, even after all these years, I'm still not sure that my ambitions for the studio were met.

Architecture can be understood as the juxtaposition between space and matter and how each can contain the spirit if not the fact of what I would refer to as uncanny. It is now six and a half decades since I made the decision to become an architect and while it is clear that matter defines space, I grudgingly admit that spatiality is not my long suit. I have never been able to get my arms around methodologies by which space (yet alone spaces that exude aura) evolves, the result being that I never developed any theory by which spaces came about in my work systematically; on the other hand, matter is my game. I adore making things or frankly even observing how "stuff" is made by others and while materiality is but one part of the architectural equation, until recently it is the part on which I have focused throughout my career.

*   *   *

In the same spirit that the post-structuralist critic Roland Barthes was repulsed by what he perceived as the artifice connected with American professional wrestling where competing athletes sometimes thrash about in feigned pain, historically speaking architecture has been for most practitioners the antithesis of artifice. However, in order to best reveal the culture that brought it to life, architecture needs to reflect both; Robert Venturi's celebrated, but selectively limiting concept of the Decorated Shed as defined in his Complexity and Contradiction in Architecture treatise notwithstanding. The theory that a building might be composed of parts that did not concomitantly signify what they were meant to convey is unimaginable to the fostering of architectural understanding, assuming that the communication of ideas to those not exposed to the arcane side of the discipline plays an important role to architects in the first place.

In Jewish mysticism, my interpretation of the third part of Lurianic Kaballah is "a failed attempt at healing an irreparable wound;" the working title of an as-yet-unpublished book on which I have labored for nearly two decades. My understanding of the Kabalistic theory underpinning a "failed attempt" is related more closely to an unresolved dialectic than to the Hegelian concept of synthesis from which I believe it to be utterly distinct.

The conceptual metaphor of "failed attempts" emanates, at least in part from that medical convention that proposes that when the slit skin of a wound is stitched back together, over time virtually all of the stitches disappear with the single exception of the last stitch which is the palimpsest of the knot that initially held all of the stitches together. In that same spirit, architecture never heals anything: architecture (not so) simply expresses the effectuation of shelter. It is the innately human need to attempt to heal that is at least as important as that which has been successfully healed. It is not that building per se solves problems, it simply makes problems visible by actualizing them.

Some might suggest that it is incumbent on architects to build simply to acknowledge that we were here and recognizing that, we tried to do something about it even if that something required nothing more than creating a part of the record of our own epoch.

Actually, architecture is quite a lot more than recording what we found during the time that we were here. Being an architect requires a proactive posture on the part of the practitioner. If what we found required that we needed to attempt to heal some imperfection that we discovered, it seems to me that it is morally necessary to assume that responsibility.

Given that the importance of moral and ethical issues in architecture have become increasingly clear to me over time, it has become eminently obvious that the discipline of architecture is perpetually challenging and should be entered into thoughtfully; perhaps even reluctantly. Because it is not my basic nature to do things that way, it never occurred to me to think about becoming an architect in that context initially but, through a number of happy, albeit misguided accidents, that's precisely how my career evolved. It took me decades to have a handle on the multifaceted depth into which I had delved. Obstinacy, or in my case bullheadedness, now and then trumps talent and because I stayed the course, the comprehensive nature of architecture gradually revealed its multivalent nature to me.

As an activist who has a reputation as one who will rush to judgment, I make numerous errors of commission. In one of my seminal errors of omission, I failed to engage in any due diligence before committing myself to a life career in architecture. Nonetheless, there were two overarching factors that ratified my decision to become an architect.

The first is that architecture offers numerous opportunities, particularly for those like me, burdened as I am with a short attention span. One can be engaged intermittently and/or simultaneously with theories, concepts, drawing, designing, building, lecturing, writing and even critiquing the work of others. Engaged in concurrently, the testing of writing versus drawing in order to verify that what was designed has great potential for fulfillment or, at the very least, understanding.

Secondly, and more significantly, architecture serves society. Being an architect is not just about self-gratification, although working in this field has that uplifting, if somewhat shortsighted potential. More enduring rewards await those who practice the field as a social art even as they pursue the profession both as a useful art as well as an aesthetically driven discipline. One needs only to spend time with museum directors, curators and registrars to understand that virtually none of them define architecture as a fine art. Apparently, if you walk on it, or piss in it, it doesn't qualify as fine art and is thus diminished in the eyes of those of who see beauty as more noble than useful. However, that limitation has never particularly hindered architects from attempting to break down the barriers between art and architecture. To put it another way, architects have over time intermittently struggled to make what is useful artful. For the outsider, the unspeakable becomes a kind of beckoning fair one.

One of the more noteworthy implications of what is artful in building is the ineffability that is somehow embedded in the mass that defines structure (one thinks of E.A. Poe and what strange things one might find embedded in a wall) or the space(s) that are the result of the disposition of mass (the notion of the uninhabitable comes to mind). The most unutterable of spaces are the ones that are meant for others (the deceased, for example).

*   *   *

Architecture is innately optimistic. Being so bold as to build is a representation of the determination to generate something from nothing. Architectural practice serves to actualize what is seen in the mind's eye as well as defining concepts by producing spatial evidence delimited by mass; it is the fundamental challenge of the architect to bring that unresolved dialectic to light. In that spirit, it goes without saying that an architect needs to be endowed with powerful convictions, since inertia in one form or another is always ready to question, indeed to stymie the work of those who would build, even if such work is conventional, but particularly in the case of ground-breaking design.

The audacity required to overcome whatever odds that posit a blockage to making something out of nothing is an important characteristic for an architect to develop, as is the will to continuously toughen one's resolve so as to be able to counter whatever opposition one may encounter. And as sure as the day is long, opposition will always arise in one or more of its unexpected guises to sidetrack the architect away from bringing his or her work to fruition.

*   *   *

After all these years, I think I have found a working definition of good architecture; it is the ability to say "no" and still get it built, by which the backbone of the architect becomes increasingly important in that quadratic equation.

Such determinist behavior, however, has an inherent downside—egocentrism, or the inordinate amount of resistance necessary to defend against whatever might stand in the way of accomplishment. Without question, I am one of those so endowed. While that characteristic has put me in relatively good stead architecturally, it has not done me a lot of good interpersonally.

Contradistinctively, an architect's ability to pay attention is equally vital lest you be accused of marching to a drumbeat that only you can hear. The former Yale Architecture School dean and peripatetic architect Charles Moore was masterful at listening to problems as they were presented to him by clients, individually and/or collectively, such that he was virtually invincible in competitions between colleagues as well as in the interview process toward the end of obtaining work through various selection processes.

He not only paid attention to what he was told, but he always included others in the decision-making process. This posture served him well; he could never be accused of being anything less than inclusive. Contradistinctively, he could also never be accused of producing anything autonomously.

*   *   *

Almost a half-century ago John Dymock Entenza, the universally respected founder and publisher of *Arts and Architecture* and later the first full-time director of the Chicago-based Graham Foundation for Advanced Studies in the Fine Arts, once introduced the great Ludwig Mies van der Rohe on the occasion of Mies receiving the Chicago Chapter of the American Institute of Architects' Gold Medal as someone "who had to will his buildings into existence." At the time, I was astonished by the audacity of that comment. Today it is my mantra, precisely because strength of conviction or will is precisely what is required to prevail over and against the inertia that often raises its ugly head to thwart accomplishment.

In 1924, the same Mies van der Rohe was quoted as saying that "architecture is the will of an epoch translated into space." Ergo, one could argue that reflecting society's will is a reasonable ethical response for an architect concerned with being responsible. Yet, equally essential is the concept that an architect be a kind of Pied Piper, so as to better lead society in directions that improve upon it; which is to say that one attribute for the architect is to be responsive to circumstances as they arise while another characteristic is contradistinctively to be proactive. In other words, mutually exclusive forces sometimes appear to exist that simultaneously seem to pull the architect in directions that appear to be diametrically opposed to each other.

From an architectural point of view, it can be argued that it is commendable, if not absolutely crucial to be able to make decisions about the appropriate use of matter. Those decisions sometimes appear to be near Solomonic in the often titanic struggle to even contend with the disposition of matter in and of itself, to say nothing of the appropriateness of what one designs for human habitation when that is factored into the quadratic equation that underpins building.

But materiality and the spaces that are defined by them, that is the tools available to the architect, also contain the seeds of the ineffable. The search for that which is generally indescribable is at the heart of the architectural project, particularly for the outsider.

*   *   *

Given all the salient requirements of a successful career that I have stipulated, it would seem that the very last thing that a person like me who seeks instant gratification, and is ostensibly not in it for the long haul, would want to become is an architect. What is it about a field that appears to be so resistive to penetration yet at the same time is so sublimely seductive that it would attract an apparent nomad? Is the grass so much greener that, whatever the effort required, there is the perception that something of a reward exists for those so inclined to scale walls to reach those verdant pastures?

*   *   *

Capriciousness has its own price. Without a signature style, one's lack of predictable architectural design resolutions can be a danger sign to even the most sophisticated of clients. For example, in 1985 the groundbreaking modernist furniture manufacturer and distributor Herman Miller commissioned Frank Gehry on Margaret's recommendation to design a major facility in Sacramento, California. The company never had second thoughts or questions about what Frank might design for them; they had researched the nature of Frank's oeuvre, however outre, and they were not only content to contend with the results, but most likely looking for a Frank Gehry signature building. It was precisely the predictable quality of his work, his signature style as it were, that justified their hiring him in the first place. However, when Frank suggested to them that as a friendly "thank you" for our recommendation he wanted me to design a small scale cameo performance piece in the complex so that he could in turn design a responsive structure to challenge what I had done, Herman Miller's resistance to that idea was conspicuous.

On the basis of what Herman Miller knew about the formal nature of my work, they couldn't predict what the result of my involvement in the project might mean. That is to say that they worried more about an unknown quantity than about an avant-garde result with which they were familiar.

Herman Miller facility, Sacramento, California, 1987

Only through Frank's determination to prevail on behalf of our friendship was he able to persuade the Herman Miller decision makers to relent, allowing him the luxury of accomplishing his intension of bringing me on board so as to create a counterplay.

1

THE CONSCIOUS EXPRESSION OF A CHECKERED CORE

My story unfolds at the onset of the Great Depression. My father, who was a late-1920s graduate of Chicago's Armour Institute, now the Illinois Institute of Technology (IIT), had acquired an undergraduate diploma in chemistry by amassing enough credits to graduate by attending the institute's night school that was located in a neo-Romanesque red stone pile on the near south side of the city. Significantly, IIT was a stone's throw from Comiskey Park, home of the Chicago White Sox, a Major League Baseball club that I was destined to support. Due to massive unemployment following the stock market crash of 1929, throughout most of the economically challenged decade of the 1930s my dad was unable to get a job in the field for which he was trained.

While he struggled vainly to establish an independent practice as a chemist by manufacturing and distributing litmus paper, my father also worked periodically for the Chicago Park District picking up debris with the business end of a sharp stick in the area adjoining the western shore of Lake Michigan in Lincoln Park. My mother earned a modest salary as a low-rated government service clerk-typist employed by the Federal Government in downtown Chicago. She felt their joint income was insufficient to provide them the security of independent living. My mother's quest for economic stability caused endless argumentation between the two of them about their dire economic straits. When their critical financial situation became further exacerbated by her one and only pregnancy, she reluctantly opted to move in with my father's parents who ran a north-side boarding house abounding with a colorful cast of characters who, like the lake itself, came in and went out.

Thus it was that my early childhood from birth at the beginning of the Great Depression through to eighth grade was spent enduring that cataclysmic decade in my paternal grandparent's three-story, wood-frame house in Chicago's Edgewater District two blocks west of Lake Michigan.

*   *   *

My grandmother was the chef at the Belden Stratford Hotel in Lincoln Park. She also moonlighted by catering weddings for wealthy Jewish hotel patrons. Jolly yet stolidly matter-of-fact, rotund Rosa Tigerman (nee Weiss) was without doubt our little family's most significant profit center.

In Europe my grandfather, Max Tigerman, had been employed as a tailor in Miskolc, Hungary. After immigrating to the United States at the beginning of the 1890s he never worked another day in his life. Instead of finding a tailoring job, my grandfather retired to the third floor memorabilia-filled attic of our house reinventing himself as a Talmudic scholar; a self-proclaimed *yeshiva bucher* whose entire life in America

Max and Rosa Tigerman family portrait, 1928

Emma and Sam Tigerman with their son Stanley, 1933

from that point forward until his death in 1938 was devoted to Torah studies interspersed with Talmudic interpretation. Without official sanction, his Jewish scholarship was outside the mainstream of those who felt it important to contribute to the well-being of his family; thus I perceived him as an outsider—an intellectual without portfolio.

These two grandparents with their disparate personas—she no-nonsense who used food as a conduit to the heart, he sometimes musingly mystical, sometimes the well-decked-out roué—had an unintended influence on the differentiated ways by which I perceived life. Without privileging either direction, I was to continuously test each position one against the other far into the future.

Most days my grandpa scrupulously studied torah simply for the pleasure of grasping its interpretive tenets. Lying in bed clad only in his soiled long johns and smoking awful-smelling cheap cigars, he alternatively labored at deciphering Talmudic exegesis. Cloistered in his attic hideaway tucked under the street-facing center gable of the third floor garret of our house, he hardly ever shared his thoughts about his studies with me but my child-like impression was that his work had a significance that transcended the mundane practicalities of life.

Most nights my grandpa reconstituted himself entirely. Dressed up in one of his snappy three-piece suits bottomed out with spats, gold pocket watch noticeably anchored to his vest pocket and a dress shirt replete with tie pin and cuff links, he cavalierly sauntered off to play pinochle, schmooze and sip schnapps with his Magyar cronies as he made the rounds of various smoke-filled Hungarian coffeehouses on Chicago's west side. To my mother's chagrin but to my endless delight, when I was six years old he began to take me with him on the streetcar to reconnoiter those outwardly disreputable establishments.

I never entirely understood how he was able to compartmentalize the disparate parts of his life so deftly, but I unquestioningly admired each side of his schizoid personality.

I was never sure why he took me with him on his forays into "the dark side" but he encouraged my budding pinochle skills, such as they were. With more love than sense, he led the family to believe that while with him I had mastered the game of pinochle at a preposterously precocious age as he gambled the night away in those dens of iniquity. Perhaps by building up my dubious card-playing capacities, he further rationalized his absences from nighttime arguments at home about financial issues, for which he had neither interest nor answers.

Groomed to the nines with his well-trimmed hair and full moustache, my grandfather was at once handsomely well-turned-out and easily enchanting. Nonetheless, a product of his middle-European generation and gender, he was without question a misogynist. During one particularly high-decibel disagreement between my grandparents—a relatively common occurrence in our house—I remember his exacerbating the argument by suggesting to my grandmother that, "a woman's place was in the kitchen with the rest of the pipes!" To say the least, I was startled to see that my grandma's knee-jerk response to her spouse's outrageously sexist remark was to chase him 'round and 'round the kitchen table brandishing a glistening butcher knife. I ducked out of sight underneath that big round slab of wood. Needless to say, ours was a highly hot-tempered domestic, if not domesticated, establishment. Hungarian Jews are anything but emotionally constrained.

I witnessed their periodic skirmishes from my many-legged hideaway beneath that generous kitchen table where I languished lazily on a cushion, most days dreaming and drawing cartoons as I observed out of the corner of my eye my grandma putting a pinch of this or that spice into some richly seasoned gravy, soup or sauce. Panic would set in when she ran out of either spicy or sweet paprika. Even now, I daydream of becoming a saucier where my concoctions would compete with my grandma's for culinary distinction.

She rarely used measurement as a guide to her kitchen innovations, relying instead upon her finely tuned intuition about all things gustatory. Her gastronomic preparations included such savory yet unbelievably greasy Magyar gourmet specialties as Hungarian goulash, chicken paprikash and apple strudel. The pungent paprika-laden aromas of that era are permanently stored in my sensory memory bank.

Today, whenever similar smells waft across my senses, I envision my amply bosomed, thickset grandma anchored solidly behind her soiled, yet unimaginably flavorful apron, and my mind reduces those many intervening decades to a heartbeat. In all these years I have yet to savor strudel dough stretched to near transparency, separated and as wafer thin and buttery as what she pulled across that well-oiled, wooden kitchen table.

*   *   *

As a counterpoint to my grandfather's self-endowed, yet not always clearly enunciated theologically based intellectual conceits, exerting absolute control over the kitchen was my grandma's way of dominating her family's always troubled domestic affairs, as she attempted to alleviate financial stress with full-flavored concoctions.

Since no meal that my grandma prepared was ever made the same way twice in succession, her immediate family as well as her boarders remained in a permanent state of apprehension at mealtime. Irrespective of what dish she prepared for dinner grandma always asked and then immediately answered her own question, "Good, isn't it?" I often thought of commenting, "I'll be the judge of that," but I was too intimidated to challenge her gustatory authority. Now of course, I consider myself charmed to have inherited a bit of her instinctive culinary skills, but throughout early childhood my home life revolved around my grandma and her warm, welcoming and wonderfully odiferous kitchen.

Other than her immigration to America, my grandma's most noteworthy adventure was that of sustenance in all of its forms from food purchases and preparation through transformation to ingestion. Seeing the tenaciously innovative way that she practiced her craft, I came to first comprehend, and then to appreciate the value of both behavioral systems.

The downside of my grandma's gastronomic supremacy in the kitchen was the lack of confidence that it wrought upon my mother's already fragile sense of self. My grandma utterly dominated her kitchen, brooking no interference from any outside source whatsoever, much to my mother's increasing chagrin as well as to her decreasing self-esteem. She looked upon herself as an outsider even within her husband's family. I'm not sure that my mother ever prepared any meals that I'm aware of. Since my mother predeceased my grandmother, I never even knew that my mother could do anything in the kitchen more complicated than boiling water, because she was never given the chance during my life with her.

While I am the cook-of-record in my household as my son JJ is in his, my grandma had a massively better track record of successful meals made and delivered than we ever had, since neither of us had to get it right or get fired. I have the distinct feeling that our success is based on our respective spouses' humor rather than their appreciation of our self-appointed roles as chefs de maison.

More importantly, and for better and worse, I made mental notes of my grandma's high-handed way of replacing measurement with her casual approach to quantities as she prepared her concoctions. My grandmother's hurly-burly way of life was never determined by regulations established by others; indeed she set no limitations on the way she went about her business either in ours or the Belden Stratford Hotel's kitchen. My own resistive path to conforming to any set of strictures in architecture, indeed in life, leads back in no small measure to my observation of the self-sufficient, independent way that my grandma approached her kitchen duties.

*   *   *

My paternal predecessors together with my mother's parents, both of whom died before I was born, were first-generation Americans. Coincidentally all four arrived in Chicago just prior to the 1893 Chicago Columbian Exposition. They all immigrated to the American heartland from eastern Hungary; my father's parents from the industrial city of Miskolc, and my mother's parents from Beregszász, now Berehove, just across the Ukranian border at the eastern edge of Hungary.

As Hungarian nationals, they all had initially migrated north from their common origin near the Borgo pass in Transylvania when that Hungarian province was threatened by Romanian expansionism. To keep me in line as a toddler, at bedtime my grandparents recounted apocryphal tales about the ficticious vampire Count Dracula and his not-so-fictional brutal counterpart Vlad the Impaler. These stories dramatized the odd person abroad at midnight in the mountains of Transylvania, with its notorious Borgo Pass, who was discovered face down and bloodless in the forest at dawn with two holes piercing that unfortunate nomad's jugular. It didn't go unnoticed by me that the safe keeping of these Transylvanian homes loomed large to those who felt vulnerable to unknown dangers. Closed shutters and doors locked the inhabitants safely away for the night. I can't say that those fearful fairy tales influenced my eventual fascination with Jewish mysticism, but they certainly didn't suppress that interest either. Those stories were told to me over 70 years ago, yet even now I still double-check doors and windows to make sure they are either secured by dead bolts or latches before I feel satisfied to retire for the night.

*   *   *

As the 19th century came to a close my four grandparents, along with many other terrified middle-European Jews, fled Eastern Europe as the dreaded, near-century-long Jewish Pale of Settlement with its notorious anti-Semitic pogroms threatened to expand westward, engulfing Austria, Hungary, Romania and the Balkans.

After the earliest partition of Poland in 1791, Czar Catherine II ("the Great") artificially established the Pale of Settlement, an enormous expanse the locus of which was 50 degrees latitude and 30 degrees

31

longitude in which Russian and Polish Jews were forced to reside. By 1880, over two-million Jews began an exodus to the United States, Great Britain, Europe, South America and Palestine principally from the town of Brody on the Eastern border of the Austro-Hungarian Empire. Collectively, they became outsiders to an outsider status bequeathed to them by Catherine the Great.

Although my grandparents' relocation to America preceded by a quarter century that portion of Hungary being over-run by Romania, given the option of becoming segregated by colonizers in the land of their birth or migrating to any democracy that would welcome them, the choice was an easy one for my predecessors. Those who had no sense of a national homeland anyway felt pressure from Eastern European nationalists whose patriotic fervor was exacerbated by centuries of latent anti-Semitism not always discouraged by their own governments. Nationalist zealots ran amok as they wreaked havoc among the Jewish populations of Eastern Europe thus instigating massive migrations to other parts of the world.

*　　*　　*

In the 1920s my parents were introduced to each other at one of the regularly scheduled Sunday outings organized by and for the rapidly growing Hungarian community. The community traditionally gathered on the west side of Chicago in one of the city's public parks in a failed attempt to find a home amongst their own kind. At that time there was a decidedly clannish Magyar constituency (is there any other kind?) that settled on the west side of Chicago and spilled over into suburban Oak Park. There, before his family relocated to the north side of Chicago, my grandfather raised his three sons to become diehard Chicago White Sox baseball fans. By taking me to Sox games well before I started attending kindergarten my grandfather made sure that those loyalties were instilled in the next generation as well.

Photo of the author submitted to a baby beauty contest, Chicago World's Fair, 1933

Growing up in my grandparent's Chicago boarding house was a gas, in no small part because its constituents were culturally, behaviorally and racially diverse and therefore of great interest to an overly-protected only child—especially that pampered "little doll" who won a beauty prize in the 1933 Chicago World's Fair: A Century of Progress. My mother was extremely proud of her child's recognition, in part I suppose because she was so frustrated by her own unfulfilled lot in life.

To suggest that being spoiled is the purview of the wealthy is an elitist myth; all you need be is an only child, particularly if you grow up in a boarding house overflowing with both permanent and transient adults who often pander to the whims of an overindulged, overstuffed little creature, who thereby comes to believe that he is entitled to be the center of everyone's attention.

Now this early enculturation of my sense of entitlement had more than a little to do with my established behavior, to my periodic chagrin both as a person, as well as an architect. I have unfortunately, on more than one occasion, informed others that they play the game with my marbles, or not at all. While this tactic may appear marginally successful in attaining some degree of sovereignty in the decision-making process, it accomplishes little in the larger scheme of things. The longer I live, the more I realize that Socratic interactive discussions by which you engage others in debate transcends professing by far as a condition by which you develop your own social skills.

Always prepared to rebel, I resisted and resented any interference that might place constraints on my natural sense of independence that was heightened by those who so enthusiastically spoiled me. Children with siblings learn the value of Socratic interaction soon enough by being subjugated to sharing with others on the relatively flat playing field of democratized family life refereed by their elders. Without pointing fingers at others, an only child in a boarding house doesn't readily come across the means and methods to give-and-take. Taking seems to be the more superior option and an only child learns that lesson soon enough.

*   *   *

During my pre-school years and through the third grade the single most significant individual in my young life was my grandfather, largely because he was the only one continuously available to me during my waking hours while both parents and my grandmother labored outside the house. Max Tigerman had a lopsided influence on my life as the only available adult with whom I had a kinship. Our relationship, though continuously loving, isolated me from other more disciplinary influences. But more than anything else, for me his retreat from the practical world, with its seemingly irresolvable challenges, into the spiritual realm of theology added a distinctive aura to his personality. This caused me to challenge pragmatic responses to problems as they surfaced and enhanced my growing interest in ineffability. I endowed my grandfather's status from the quirkiness of the outsider to a position of privilege that his scholarship somehow benefited him.

When I was at a precocious age, my grandpa taught me to speak and to read both English and Hebrew, going so far as to enroll me in the pre-kindergarten program in the nearest orthodox Hebrew school, the *shul* at Agudas Achim. After my grandfather's death, when I was nine years old, my mother ambitiously re-enrolled me in Chicago's up-scale north side Reformed Congregation at Temple Sholom, whose membership included her wealthy brother Samuel Stern.

With its Middle Eastern octagonal stone-clad prominence on Lake Shore Drive, Temple Sholom was so reformed that it scheduled Sunday services with minimal Hebrew in lieu of Friday night prayer for its largely upper-middle-class congregation. We smart-ass kids referred to it sarcastically as "the church on the lake," thus modifying its official sub-title from "the temple on the lake." Being virtually the poorest kid at Temple Sholom reinforced my *auslander* status even as it caused me to question my sense of belonging to a religious institution whose perks were, on the face of it, seemingly devoted to the moneyed constituents of its congregation.

*　　*　　*

While my grandfather supervised my first steps and showed me how to walk, more often than not he entertained me with tall tales about our family origins. Once, when I was five, while the two of us were rummaging through memorabilia stored willy-nilly in the attic of our house immediately aft from my grandpa's lair, we stumbled across a colossal oil painting that depicted an armed great white hunter squaring off against an enormous Royal Bengal tiger. When I asked him if the painting had anything to do with the Tigerman family name, he responded with a possibly apocryphal story about one of our ancestors, a Jewish serf in Hungary who lacked a proper surname. While serving in the Austro-Hungarian army during a battle against the Turks, this soldier volunteered to go on a mission behind enemy lines.

According to my grandfather, our ancestor accomplished his assigned military task with such remarkable skill that his supervisors in the Austro-Hungarian military leadership rewarded him with the incomparable gift of a last name whose meaning was appropriate to the nature of his daring achievement. Thus, he was named "tiger" for his courage and "man" for his intelligence. In Hungary our family was first known as Tihany, then Tiegermann and finally, thanks to the renaming authorities at Ellis Island, Tigerman. The impact of that story recounted to a wide-eyed five-year-old by his beloved grandpa left an indelible impression still vivid after all these years.

Little did I know then that some three decades later I would find myself stalking a Royal Bengal tiger on foot as our hunting party of five padded cautiously through the moist alluvial clay of the steaming jungle of the Sunderbans, a beautiful forest south of Dacca and east of Calcutta in what was then East Pakistan.

My grandpa was my baby sitter, my playmate, my tutor and my personal cultural resource and I loved him with all my heart, but one summer afternoon his self-imposed isolationist lifestyle caused me the panic of my young life. Playing in the front yard of my friend Bob Schwartz's apartment building just down the block, the two of us were excited to see a big red fire engine, its bells clanging and its siren wailing, racing down the street in the direction of my house where it came to an abrupt stop at the curb. We anxiously chased it down to discover smoke billowing from the open third floor double-hung window of our house. Clad in his long johns and still in bed my grandfather, with a smoldering stogie still stolidly clutched between his fingers, was being carried out of the window much like a tray of hors d'oeuvres by two firemen. Apparently, the old man had fallen asleep while he was engaged in Talmudic exegesis. Anyone who has ever attempted this tedious task understands the likelihood of his dozing off. As he smoked his cigar he most likely caused the bedding to first smolder, then to catch fire; a vivid, still lingering image.

In my childish eyes, my grandfather was larger than life. He could do no wrong, even though I was terrified late one afternoon in that stuffy attic when he killed (what I thought to be at the time) a large spider with the palm of his bare hand by squashing it to death as it crawled down the king post that I was leaning against. I understood that he was protecting me but I have been apprehensive about spiders ever since.

Since I was to spend a generous portion of my future life in East Pakistan where spiders the size of dinner plates abounded, this was an ill-fated omen. The native Bengali population protects spiders because the eight-legged creatures dispense with lizards and other vermin that can be a real hazard to health. My fear of spiders only elicited a humorous reaction from Bengalis while reinforcing their observation that Caucasians are not really in tune with nature—in my case an unerringly accurate description.

My grandpa and I were together constantly. As I have already noted, he was a role model of some consequence, and if he had lived somewhat longer than he did there is the distinct likelihood that my life would have taken an entirely different turn. Conceivably, I can imagine studying to become a rebbe. But, would I have had the staying power to focus on a discipline as demanding as religion? Given my notoriously short attention span combined with excruciating memories of difficult Hebrew lessons at Agudas Achim where knuckles bloodied by baton-wielding rabbis were a daily occurrence, I have serious doubts about my ability to prepare for such a noble calling with its attendant discipline.

Nonetheless, my search for the ineffable began with my grandfather's interest in the rigors of theological interpretation in which he engaged for nearly a half-century. More than another half-century would pass before I was able to adequately acknowledge the depth of my feelings for my grandfather and understood the importance of his presence in my early life. When he died in my eighth year, his open coffin was displayed with his head facing east (in anticipation of a messianic age) on axis with the eastern oriented street-facing bay window of the living room of our boarding house. His passing represented my first exposure to death and the traumatic image of that beloved unresponsive figure became tattooed permanently onto my aching heart. I began to stammer uncontrollably within weeks of his death.

In 1988 I dedicated *The Architecture of Exile* to him. The inspiration for the manuscript was derived, at least in part, from observing his commitment to Torah. That devotion later challenged me to attempt to search for linkages between the interpretive tenets of Judaism and an architectural system that is traditionally rooted in faith and that mysteriously emanates from multivalent origins rarely plumbed by architectural historians and theorists who more normatively seek antecedent connections in Hellenic/Christian precedent. What I failingly searched for was a fusion between faith and interpretation even if I had to invent a way in from the outside that was, in a sense, a way home.

*   *   *

My grandparent's boarding house provided endless grist for my imaginative mill. The colorful characteristics of its diverse constituency proved endlessly entertaining, yet thinking back about them, I also remember their frailness and humanity. In these surroundings I was exposed prematurely to a naked level of poignant cultural diversity rarely offered to cloistered, over-protected middle-class Jewish children.

Al, who was small, secretive and sly, and Bud, who was outsized, slow-witted and warm-hearted, were the two very different African American males who lived there permanently. Bud lived in a makeshift room in the cluttered basement and Al resided in a small space behind the kitchen at the rear of the house. Reputedly, they labored at itinerant odd jobs for unidentified employers, but I'm not sure that they worked at all other than appearing to do chores around the house trading services for room and board. Of course, it was the Depression and Al and Bud's residing with us was not an uncommon occurrence in homes of caring families who helped each other as best they could during that genuinely stressful era. In our own case, the financial tension of that decade was ameliorated mightily, if infrequently, by care packages of surplus fancy hotel food that my grandma would sometimes bring home from her chef's job.

My uncle, Milton Tigerman—my father's preening, self-involved, gadfly-like younger brother—considered himself a bon vivant and at least in his own well-enunciated view, a great catch. He lived at our house at irregular intervals before his marriage to the elegantly statuesque and classy Rochelle, who to my way of thinking was entirely undeserved by him. But as a bachelor, Milton often brought home a whole host of worldly young women of doubtful provenance; at least according to my parents who hovered over me while I was in their presence lest I become exposed to aspects of life that they considered prematurely prurient.

My parents also cautioned me in no uncertain terms to steer clear of Harriet, another resident who was the reputed Jewish call girl living there as inconspicuously as she could in her small rented room in the middle of the second floor. My parents and grandparents occupied the two ample, adjacent, street-facing front bedrooms on the second floor.

This colorful cast of characters constituted the roll call of more-or-less permanent residents. Irregular short-term itinerant boarders occasionally rented one of the smaller rear bedrooms on the second floor that faced the elevated tracks. Their presence seemed to enhance the flavor of the place much like my grandma's pinch of this-or-that spice that embellished her cookery in unpredictable ways.

*   *   *

My interest in art and design began in kindergarten in an unplanned way, when I forged close friendships with three somewhat out-of-the-ordinary classmates, one of whom, Impy, drew the most imaginative cartoons. Impy's twin brother Mike, a boy named Alan and I were soon drawn into Impy's web of highly stimulating sketches.

The twins lived in the stylish Edgewater Beach Apartments facing Lake Michigan where the four of us, lying on their plush carpeted living room floor after school, created drawings together until Impy's father chastised us for "wasting time." Impy's parents were neighbors of a family whose son was to grow up to be the architect Hugh Newell Jacobsen. Hugh and I went to Swift School, Senn High School and Yale, though Hugh always good-naturedly reminds me that he has an "e" not an "o" at the end of his name and that his gentile fraternity was not particularly collegially inclined towards we lesser beings.

In the late 1930s, as we became conscious of the war now raging in Europe, the four of us produced a plethora of cartoon-like bellicose drawings. We took sides by producing animated images of armed soldiers, tanks and fighter planes doing battle against each other so as to wreak havoc on whomsoever the cartoon characters came up against.

In the latter part of the decade, like so many of our generation, we were in thrall to Adolph Hitler's Teutonic radio rants sponsored by the right wing publisher and Yale's own Robert R. McCormick and the Chicago Tribune owned WGN radio station. While these broadcasts remained untranslated except for abbreviated English summaries, we found them scary, yet strangely beguiling. This influential demagogue exploited his voice as a bludgeon, which brought forth a particularly powerful range of emotional responses from those four little listeners in far away Chicago.

Much later, I also became enraptured by the voice of the evangelical minister Robert Schuller, of Crystal Cathedral fame, who had vocal similarities to what I listened to on WGN. As I attended the dedication of that extraordinary structure in Garden Grove, California, which was designed by Philip Johnson in 1980 to act as a framework for Reverend Schuler's mellifluous opening commencement sermon, the memories of my childhood reverberated across three decades.

Perhaps the underlying reason that those cartoons carried such great weight was that I could produce them without fear of reprisal from the other three boys. Listening to Herr Hitler's diatribes, we concentrated so hard on what we drew that there didn't seem to be much need for conversation. That suited me just fine since, from the time of my grandfather's passing onward through the remainder of grade school, I stuttered so dramatically that I avoided eye contact with any of my teachers, terrified that I might be called upon to speak in class. Much of what I learned in those early painful years came about not in the classroom but in Swift School's corridors where I could talk to my teacher confidentially, without fear of the embarrassment of being called upon in the presence of others who, if unchallenged, might bring my uncontrollable malady into sharp relief with the attendant misery brought on by their mockery. It seemed that every which way I turned my outsider status was there to remind me that I was not a part of whatever collective I might otherwise have vanished into.

My stuttering along with much of the girth that I gained while eating my way through my grandma's aromatic kitchen evaporated without explanation during adolescence, but I never forgot how awkward it was to wait for the teacher to exit the classroom so I could steal some one-on-one private time with her. From my kindergarten teacher Mrs. Hook through to my eighth grade teacher Mrs. Antoine, all of my teachers were strong yet perceptive females who were suitably sympathetic about my stuttering. Those difficult school years were to have an unexpected impact on certain aspects of my architectural production many decades later where I found myself focusing on exaggerated circulation systems as a kind of approach-avoidance replacement of destination spaces. Remembering how important the passageways of my childhood were, I would endow corridors, galleries and rooms "en suite" with an idiosyncratic kind of meaning of greater consequence than that of simply providing access to the destination spaces that one might otherwise think they were designed to serve.

As a result, I became preoccupied with the concepts posited by the theologically inclined Danish philosopher Søren Kierkegaard. Perusing the theories that he expressed "in the process of becoming," I developed an intense interest in expanded circulation systems in and of themselves rather than where they led. In other words, focusing on process, rather than on product, seemed to be a way of avoiding coming to grips with closure from an architectural point of view.

I couldn't articulate it at the time, but subliminally I think I was in the process of instinctively rejecting ideality with its Hellenic-Christian origins, which seemed to be represented by finitude; closure of the sort that expressed faith in ideals rather than expressing the process by which concepts evolved. I didn't have a sufficiently developed language then to clearly express these thoughts in words: that level of informed expression came later. At a sub-conscious level, what I think I was really rejecting were the ideals representative of Hellenic-Christian architectural origins. To this day, I persist in interpreting rather than having faith in any particular set of rules or regulations that would define or, by extension, constrain architecture.

*   *   *

While history and Viollet-le-Duc teaches us that architecture begins with the "parting, or the shrinking of the walls," as in the introduction of the column in combination with resultantly elongated beams, I would submit that the pre-eminence that I would later give to corridors and galleries is well within the patrician traditions, that for eons, have defined the architecture of what can only be referred to as abundance. In the sense of profusion, corridors and galleries were also about the notion of a kind of processional that lead towards the dead-end rooms representing destination spaces that by their very placement symbolized fulfillment.

In contradistinction to the kind of civilized living available to the landed gentry represented by *en fillade* rooms was the "shotgun house" begun in America's pre-Revolutionary War era. Those notorious southern slave quarters were a study in minimizing space even as they diminished those persons who occupied them. No corridors existed at all and every one of the inhabitants could be dispensed with in one ugly shotgun blast delivered from the front door, thus its name. In the case of the shotgun house, rooms rather than corridors were en suite and there was neither preparation nor ultimately pinnacle; only endlessly unfulfilled movement back and forth bereft of the luxury of privacy.

*   *   *

The fabrication of those long ago drawings, together with membership first in the Cub Scouts, and then in the Boy Scouts, and a brief stint in a gang called "Gutland," were my major childhood activities. I was "Joe Gut" and my opposite number in the other gang, "Tough Herman Land," was a kid understandably called

39

"Tough Herman." It wasn't long before my participation in that organization was abruptly curtailed by my mother. These efforts constituted my major extra-curricular activities throughout grade school as Europe and Asia initially and then the United States successively devolved into war.

The Boy Scout troop that I belonged to, which operated out of the Temple Sholom gymnasium in those years, was led by a scoutmaster who we all referred to as "the last Jewish fascist." For example, if the troop didn't perform in an exemplary fashion in regional first-aid meets, the scoutmaster exiled us to northern Michigan over spring vacation to plant fir saplings from dawn 'til dusk on long, frigid weekends. At our summer camp on Big Blue Lake north of Whitehall, Michigan any of us who couldn't master swimming in the customary chronology of merit badge achievement were thrown in the deep end of the pier with no adult in attendance. It was literally "sink or swim." On one particular parents' weekend when I had been selected as the camp's chief cook, a fly was discovered swimming in the gelatinous dessert that I had supervised. On the spot I was abruptly and very publicly demoted from chief cook to lowly dishwasher.

If the Boy Scouts of America were dubiously thought of as a paramilitary organization, surely our overachieving scoutmaster won the award as its marine master drill sergeant. Nonetheless, he bullied those of us who survived his badgering into the rarified rapture of Eagle Scout with three palms as well as cajoling us into conquering the wilderness to be eligible for membership in the Order of the Arrow.

*   *   *

The composite cartoons that the four of us concocted that reflected our joint imagination regrettably went the way of most artifacts produced at an early age—they simply vanished.

Thirty years would pass before hybridized descendants of those cartoon characters would resurface in a somewhat different iteration. They reappeared while my two pre-teen children JJ and Tracy, my second wife JoAnn and I were in residence at Oxbow, the Art Institute of Chicago's summer colony in studio art that is situated in the woods on a quiet lagoon emptying into Lake Michigan in Douglas, Michigan, 140 miles around the lake from Chicago. Those reinvented creatures from my childhood ushered in a new, albeit ironic stage in my already multi-directional—or unfocused—architectural career.

Oxbow encouraged experimentation. While I was there allegedly to teach an architecture appreciation course, I was also charmed by the relaxed atmosphere of a summer art colony with its unexpected delights. We all took art classes and lived in ramshackle, leaking wood cabins, every one of which was replete with its own covered porch and unbalanced rocking chairs as we ended each halcyon day by dressing down for dinner. Led by the eccentrically persuasive art dealer, Phyllis Kind, Oxbow provided our little family with several seminal summers soaking in the ambience provided by painters and sculptors lit by the hot Michigan sun, during which time my soldiers and angels reemerged following a 30 year hiatus.

*   *   *

After grammar school, my twin cartoonist comrades moved to Half Day, Illinois, a small town 35 miles northwest of Chicago, and I only reconnected with them after high school when they matriculated at Harvard, in Cambridge where I made an unexpectedly abbreviated year-long appearance at MIT. The third boy, Alan Rose, ended up at Columbia University where he ultimately received his PhD in American Studies, subsequently teaching English literature first at the US Naval Academy in Annapolis, Maryland and then settling in for the rest of his time in academia at a small community college in Appalachia. The twins finished up at Harvard and Impy, aka James Kenneth Kettlewell, stayed on for a PhD in Art History.

Impy spent his entire teaching career at Skidmore College in Saratoga Springs, New York, where he became the "Mr. Chips" of the art history department. Kettlewell's area of expertise had a late-19th-century focus with more than a passing interest in Frank Lloyd Wright. After I invited Impy to lecture on FLW to the Chicago Architectural Club in 1993, he reciprocated by asking me to give his college's keynote address on the occasion of his retirement. I recall with satisfaction the fin de siecle-like nostalgia that I felt that spring night in Saratoga Springs virtually a half-century after Impy and I first collided in a north-side Chicago kindergarten.

The impact of the production of those early drawings, collectively conceived with three other like-minded pre-adolescents, remains with me as the primary impetus that first pointed me towards an art/design-oriented career of which architecture soon became the most obvious playing field that I felt I needed to cultivate.

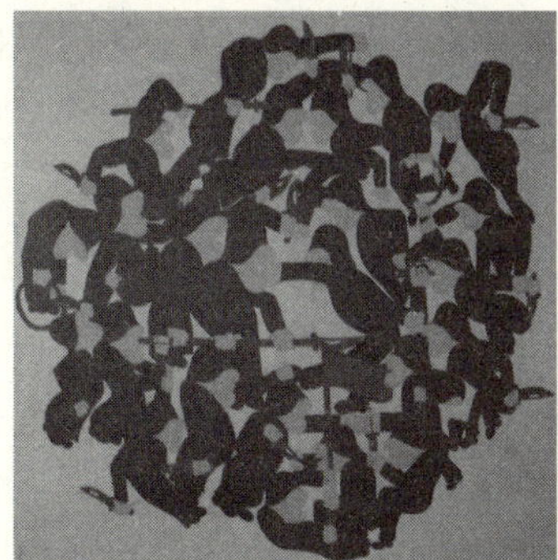

A painting by the author, 1972

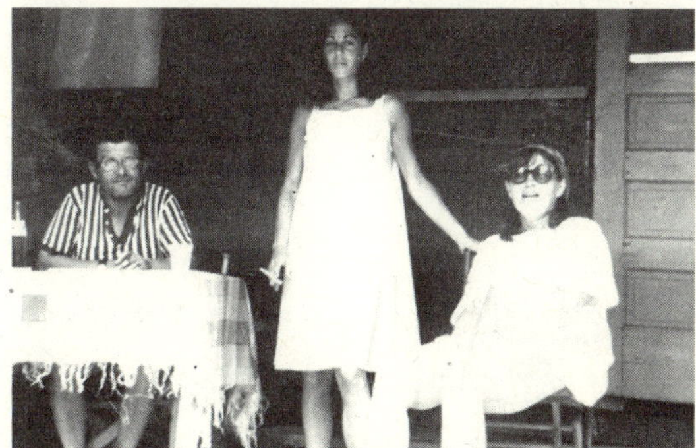

The author, second wife JoAnn and the painter
Ellen Lanyon at Oxbow, 1972

2

AN UNCERTAIN GENE POOL

On my 12th birthday, my best friend Mitzi, a chubby mixed-breed fox terrier (with the solid flesh density of a Vietnamese pig) met an unexpectedly violent end. Both Mitzi and I were born in 1930, and she was my grandpa Max's gift to me on the occasion of our joint birth. We had been inseparable for more than a decade. As I watched helplessly from the sidewalk where we had been playing, she was mortally maimed by a hit-and-run driver when she ran from the sidewalk into the street in front of my grandparent's boarding house. We had no choice but to put her down at Chicago's privately supported animal shelter, the Anti-Cruelty Society.

Thirty-five years later, I was retained as the architect for a major addition to the Society's original art-deco-like building. The intervening decades did not erase my becoming unsettled by acts of euthanasia that I had witnessed initially in 1942 with Mitzi and later in 1977 while I was being escorted through the building during the interview process.

On the latter occasion I observed with horror a "multi-tasking" teenage volunteer nibbling on a meat sandwich with one hand as she simultaneously held tightly onto a trembling white puppy while, with the other hand, preparing to inject sodium thiopental into that hapless creature's rump. Traumatic observations such as that one led me to create an architectural sketch/cartoon of one of my early designs in which I depicted my addition to the Anti-Cruelty Society in a surrealistic mode.

The director of that facility was not amused by my politically incorrect statement. He was more interested in my creating a user-friendly addition to the original structure so as to increase adoptions. Not only did I not quarrel with him on the subject, but his predilection led me to evolve a metaphorically-related façade akin to a basset hound's image. In any case, the playful façade cladding the finished building barely masks the image that I had depicted in that early cartoon.

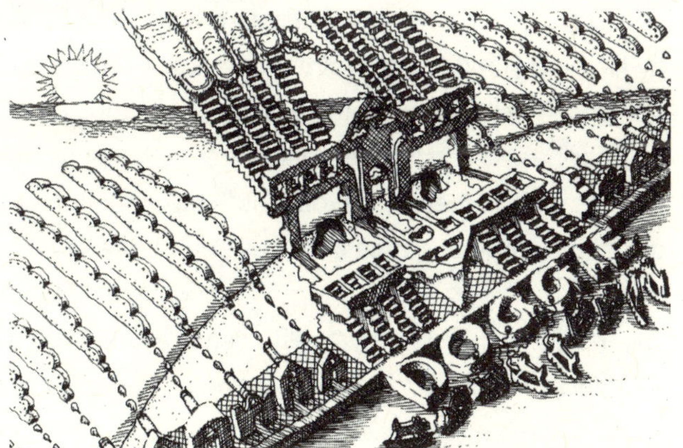

Anti-Cruelty "architoon," 1977

*   *   *

In 1943, and for no particular reason beyond unfocused recreational reading, I first scanned and then, with prurient eagerness immediately reread Ayn Rand's notorious best-selling book *The Fountainhead*. Initially published in 1943, that first-edition copy with its gratuitous sex scenes on pages 229 and 301 is still on the bookshelf in the architectural section of my home library. As a direct result of the powerful impact Howard Roark's persona had upon my pubescent imagination, I decided then and there to become an architect. My spur of the moment decision can be perceived as shallow when seen in hindsight, but overcome by yearning albeit for something I didn't entirely understand, it didn't seem that way at the time.

Certainly, I didn't resemble Ayn Rand's central character either in body or in spirit, but by identifying with Roark, I was at once rebelling subconsciously against my mother's desire to attain what she wasn't otherwise able to possess as well as rationalizing my outsider position. My mother equated success with money, both of which eluded her throughout her frustrated, unfulfilled life. Perversely, and as I believe Ayn Rand intended, I interpreted the protagonist in her book in his rebellion against our conventional understanding of success as being heroic as the author endowed him with reasons to justify his outsider status. That balloon was to be pricked by the author herself two decades hence when I unfortunately came across Ms. Rand at Yale.

Over time, my antagonism towards the status quo can be traced to my rebellious nature in childhood, in which *The Fountainhead* played a brief yet decisive role. Reading Ms. Rand's book I felt empowered by Howard Roark's determination to buck the system, the consequences of such behavior be damned.

While I still identify with the underlying message delivered in the book, over the past 68 years much has happened that has modified my initial simplistic response. But my desire to stand alone, neither influenced by others nor typecast, has remained as perhaps the overarching way in which I have made decisions, large and small, for better and often for worse, throughout time.

*   *   *

Thus it was that in my twelfth year I applied prematurely to the Massachusetts Institute of Technology (MIT) for guidance as to what was required to ultimately gain admission into its architecture program. I vaguely recall that someone in the family had contacted the American Institute of Architects (AIA) who in turn advised us that in their view MIT was the best American school of architecture at the time. Even if I had known that Louis Skidmore, Gordon Bunshaft and Walter Netsch had graduated from that institution and that, more or less, they would figure prominently in my later life, it wouldn't have meant anything to me at that young age.

Mind you, it was 1943 and I was 12 years old when I initiated correspondence with MIT, but this anecdote reveals a consistent characteristic of mine—acting without thoroughly thinking through the consequences. I play chess much the same way.

And just like that, the die was cast. No obvious evidence of hidden talent, no insight into what such a simplistically arrived at decision might ultimately entail, and certainly no due diligence of a self-comprehending nature could have dissuaded me from my recklessly made life-altering decision.

It's not as if there were other architects in my family that encouraged me to read *The Fountainhead*, although more than a few of my relatives had a modicum of artistic talent. My mother's distant cousin, Aubrey Kellner, was a moderately successful portrait painter who painted my portrait in 1939; other relatives were involved in music or theatre. Much later, my father's first cousin Richard Tigerman became the cultural director for the city of Monterrey, California, while my second cousin Jimmy Tigerman became a sought-after interior designer and art gallery dealer in Chicago.

Portrait of the author by
Aubrey Kellner, 1939

Mostly, however, my family was composed of lower-middle-class working people. My father's intimidating older brother, Joseph Tigerman, was a tough-talking Chicago beat cop who also happened to be Loyola University's track coach, the trainer for the United States Olympic race-walking team and an advisor, colleague and lifelong friend of the controversial long-time head of the United States Olympic Committee, Avery Brundage. Joseph's daughter, my cousin Sue, was my childhood playmate and the relative to whom I was closest both in age and in temperament.

As a youngster, from time to time I rode the streetcar to the Chicago Stockyards to visit a favorite family member of mine, my mother's older brother, Herman Stern. His responsibility as a supervisor at Swift and Company was to oversee the slaughtering of cattle. Many decades later, I would give a slaughterhouse as a studio design problem for student architects at Yale.

While my dad, Samuel Judea Bernard Tigerman, the gentle pompadour-topped middle son proffered the carrot, my mother, Emma Louise Tigerman (nee Stern), wielded the stick. Early on she appropriated the role of family disciplinarian and I had every reason to pay attention to her mandates of the moment. Never having any friends who lasted throughout her troubled life, I was the focus of her every whim and

PROJECTS' "Sam and Emma" cookie jar, 2000

will—I was her alter ego and my street smarts can be traced back to her. More than a half-century later, I memorialized them both when I light-heartedly designed a Janus-figured cookie jar entitled Sam and Emma that, in greater likelihood, should have probably been dedicated to my grandmother.

Thus it was that the female of the species would most influence me. My mother, grandmother and all my grade-school teachers were by and large strong-willed women. As a result, I have always been attracted to, and never intimidated by, women who have profound levels of determination.

During the school week after class and on Saturdays, while my pals were on Swift School's playground playing softball, my mother reinforced my outsider status by insisting that before I did my homework I practice classical piano one hour every day. This was more challenging than it might otherwise seem since my parents couldn't afford a piano. My daily practice sessions were accomplished with my fingers trying to find traction on a glossy cardboard keyboard lying noiselessly on our kitchen table while I pounded away in silence, at least when my grandma's rolling pin wasn't stretching strudel dough to its limit. The only time I actually heard what I played was on Sunday when I rode the bus to my piano teacher's studio for my weekly piano lesson. Rather than instilling discipline, those many hours of unresponsive practice triggered a life-long insubordination against any particular routine that was imposed upon me by anyone.

Shortly after my grandfather's death while noting her eight-year-old son's lack of concentration, my mother decreed that, "I was destined to become a jack-of-all-trades, but a master of none." Her prophesy might have been precipitated by a mother's awareness of her son's short attention span combined with the frustration of finding herself entirely incapable of modifying that behavior.

More than most pronouncements that emanated from her lips, this one adhered to me like a cheap suit. Four years later when I made the decision to go into architecture, I didn't realize that my short attention span would manifest itself in ways that would go against the very grain of normative architectural traditions. As my scattershot characteristics became progressively more unmistakable, my mother's earlier declaration came to haunt me in my later professional life.

While there are isolated antecedents in art and architecture representing multivalence as a signification of seriousness of purpose, they are few and far between. In modern architecture as their styles changed from time to time, Frank Lloyd Wright and to a somewhat lesser degree Le Corbusier would seem to qualify for that position. In modern art, it is generally accepted that Pablo Picasso is the prototype for passionate erratic behavior in his art as well as in his life.

Nonetheless, the vast majority of architects whose work over time is categorized by architectural historians as unforgettable engage in what can be most clearly described as signature work of an overarching monomaniacal purpose and form. Historically, one thinks of Karl Friedrich Schinkel and Andrea Palladio, and much later, architects that I knew. Ludwig Mies van der Rohe and Paul Rudolph are examples of architects with singularly focused careers, at least from a formalist, and in Mies's case, a constructional as well as a structural point of view.   Contemporaneously, Peter Eisenman, Frank Gehry, Charles Gwathmey, Michael Graves, Richard Meier and Robert A.M. Stern are a few among many of my peers whose self-similar signature work suggests focused like-mindedness, differential stylistic considerations notwithstanding. It is important, I think, to note that the national origin of many of these latter-day "star" architects is American. In the most free-based capitalist society of all time, is one to then assume that American architects who produced signature work were proto-branding exemplars?

Needless to say, signature work provides clients as well as critics a familiarity with one's architectural production. This is not an insignificant characteristic if one wishes to attract clients on the one hand or to be front and center on the minds of architectural critics on the other.

Furthermore, such an approach appears to legitimize an architect through identification with the thought that all there is to be learned can be summed up by understanding a single drop of water. Over time, many have weighed in on the drop-of-water theory. For example, Mohandas K. Gandhi once said that, "The oceans are composed of drops of water; each drop is an entity and yet it is a part of the whole." In the 20th century, Ludwig Mies van der Rohe more than anyone else was the architectural archetype for such focused intensity that produced an unparalleled body of apparently self-similar work.

Counterintuitively, in many instances a short attention span like mine can be extremely useful in a field normally suited to focused folk. I never felt compelled to develop a signature and thus thought it beside the point to repeat myself. I am not assigning value to this method of producing self-similar structures; it is simply a critical self-assessment. By never having to continuously function within the confines of any particular palette, I have always been endowed with a sense of artistic freedom. Nonetheless, designing with a tabula rasa mentality may be refreshing but it is anything but reassuring if not terrifying. Yet in my case, such an approach seemed to work well in the context of my short attention span.

Architectural influences—mentors and models

Reinforcing my ingrained need to constantly create anew, a number of different architects with divergent views one from another, and with whom I had far more than casual contact, influenced the several idiosyncratic ways in which I came to grips with understanding this discipline and then expressing those interests in my own way. Each of these architects approached their own work distinctively, and the field of architecture in general very differently. To choose between them engendered conflict. If there is an overarching theme in what follows, it is embedded within the architecturally uncommon subject of conflict in and of itself—a fitting response from an outsider.

In the sense that Henri Poincare's 1890 non-periodic orbits were later developed by mathematicians as ergodic theory, chaos theory opened up a new way of challenging mathematical understanding. Similarly, unresolved conflict provided me with an idiosyncratic way of perception bereft of any obligation to resolve whatever it was that I was analyzing at the time.

*   *   *

The first architect to influence my thinking, the one to whom I initially apprenticed was the free-spirited, non-conformist George Fred Keck who, like Frank Lloyd Wright before him, hailed originally from the hills of Wisconsin. I was 18 years old when I joined his eight-man office on the sixth floor of a nondescript office building on Chicago's famous Magnificent Mile.

Keck made it eminently clear to me at the outset of my apprenticeship with this curmudgeon that the very act of designing buildings was grounded in technique equally inspired by the science as well as the art of architecture. He insisted that before I attempt to imagine architectural form in any abstract sense, I needed to understand the implications of what such a concept might mean materially as well as what I wished it to become conceptually. In other words, he stressed the importance of actualization as having a constructional component that defined it, that perhaps even transcended an aesthetic or even a useful one.

Thus, from the beginning of my exposure to building, I was led to believe that architecture was definitively distinct from art. More than anything else, I was persuaded that this "useful art" needed to behave believably since the field was inextricably cultivated by its materialized context. In so doing, understanding—to say nothing of governing—the building's making was vital to its performance. I was inordinately influenced by Keck's interest in the material component, not the spatial qualities, of architecture.

The crusty Keck also insisted that being casual about drawing led to being careless about building. For him, neither behavior was even minimally acceptable; if you compromised a drawing that produced something less than what he would consider ideal, by extension you

George Fred Keck, 1949

would predictably compromise a building and such behavior was basically out of the question to his way of thinking.

I didn't quite understand it at the time, but Keck's focus on the science of building precluded any explanations he might have offered about "aura." I'm not sure what such a clarification might have included since I never felt that poetry per se was a part of Keck's vocabulary. While his architectural production was meticulous both in process as well as the resulting output, he never claimed that ineffability was something that he sought or frankly particularly cared about.

Nonetheless, Keck had an enormous impact on my understanding about architecture. The methodical way that he went about the business of what is now referred to as "sustainable architecture" was eye opening. For a teenager, I was utterly in thrall to his interpretation of reductive modernism.

*   *   *

An enormous influence in my understanding of architecture and a role model of some consequence to me and to others who came in contact with him was the resolute 20th century master builder, the German-born Ludwig Mies van der Rohe. Mies exposed me to unresolved dialectical philosophies, such as faith versus interpretation, that through their unexpected juxtaposition opened up astonishing architectural possibilities that might plausibly lead to inconclusively derived forms. This is not an insignificant discovery in a field whose normative understanding about building is grounded in fusion and rooted in faith all leading to our perception that structures are most persuasive when they are holistically derived.

Ludwig Mies van der Rohe, 1959

My interpretation of the ambivalence that I'm alluding to in Mies' work becomes evident when situating one's self at a particular distance while observing a given façade of any one of his tall buildings such that the edges are beyond the limits of one's peripheral vision. Much like observing a Mark Rothko color field painting, when the painting's perimeter is beyond one's range of vision, something metaphysical seems to mesmerize the viewer. In Mies' case, beautifully proportioned building façades (utilizing the golden mean proportional system) reductively composed of mullions subdividing glass without either beginning or end take on an almost mythic vision.

When I became aware of the inexplicable way that Mies' façades impacted my perceptions about architecture, I began to get an understanding about the power that ineffability can have on building. It was a revelation that added a dimension about possibilities connected with the enigmatic in a discipline that until that time I thought was rooted in usefulness alone. Those who think of Mies simplistically refer to his tall structures as glass boxes, however nothing could be further from the truth.

While consulting with Mies' office in the mid 1960s, I observed that St. Augustine's influential tome *City of God*, and St. Thomas Aquinas's Treatise on *God and Treatise on Man* (two of the 30 books that the SS allegedly allowed Mies to take with him from his 3,000 book library in Berlin when he departed Germany for the United States in 1937), were adjacent to each other on Mies's shelf behind his large wood desk at his 230 East Ohio Street office in near-north Chicago. I sensed that this adjacency was not a coincidence. It seemed that books on faith versus interpretation were textual neighbors in Mies' life as an architect. The significance of their placement I would come to understand better in later years but their presence suggested that the way that Mies approached the discipline of architecture was not limited to aesthetics. Clearly, his work was the result of his reflecting about philosophy and theology as well as connoisseurship.

While 5th-century St. Augustine's writings included a belief that Jews were "rebels against God," I didn't yet have language to organize questions about the linkage between Christian theology and architecture. My grandfather's heritage would become clear to me when I wrote *The Architecture of Exile*.

In a way, my attempt to coalesce Judaism and architecture represented my challenge to faith-qua-faith even as it reified the Jewish tradition of interpretation or excegesis. Writing that book further clarified my feelings about my "outsider" status that would grow throughout my life as a Jew in architecture.

But lest one think that my fascination with colliding elements was autonomous to me alone, Mies, as in so many other spheres of influence, predated my concerns about such things.

*   *   *

Much has been written about Mies van der Rohe's use of the quirk miter, a re-entrant corner detail that terminated his studiously resolved, carefully proportioned building façades causing them to be perceived almost autonomously from the next elevation around the corner. This corner-turning device that evolved in his work after he came to the United States can be interpreted as a vestigial element of walls that slipped by each other—exemplified by the 1929 German Pavilion in Barcelona, Spain. This particular resolution of the building corner resulted, at least in part from his interpretation of the ground-breaking Wasmuth exhibition that took place in Berlin in 1910 and the resulting publication of Frank Lloyd Wright's architectural production.

That exhibition displayed convincingly how Wright "broke out of the box" by setting free the corner of two adjoining perpendicularly disposed planar building façades. That strategy of transforming volumes into planes had a major impact on important European architects who attended that seminal show. Up until that time, classical building had reinforced a building's corners that "quoined" the buildings corners by anchoring large, dressed stones thus reinforcing the "x" and "y" axes of rectangular parallelepipeds. The architects of the Dutch movement De Stijl, the German architects who were instrumental in the founding of the Bauhaus in Dessau and, for my purposes here, Ludwig Mies van der Rohe were influenced enormously by what they saw in that seminal Berlin exhibition. One could argue that the Aachen-born Mies developed that aspect of Wright's effort more comprehensively in his European work than any of his other European colleagues.

Those sycophants trained at IIT by the former Bauhaus director and his earlier colleagues Ludwig Hilbersheimer and Walter Peterhans believed that Mies was exclusively modernist in intentionality. Clearly, the design of Mies' buildings after he came to the United States showed that he knew more about how the Baroque builders in Rome turned corners on their structures than his Chicago-based followers ever wanted to give him credit for. Such information might have confounded their one-dimensional beliefs about Mies's influential role in the "Modern Movement" at the exclusion of all other influences. God forbid those ideas might have originated in earlier, classically inspired work that was the antithesis of what Frank Lloyd Wright exhibited in Berlin in the first decade of the 20th century.

*   *   *

One result of Mies van der Rohe's philosophical understanding of matter and its demands was that it seemed to give him the confidence to define architecture seemingly simplistically. In the early 1950s when I heard Mies's throwaway statement delivered in his stoic style to an ambitious young ingénue/ architect's question about how this young man could better himself that, "First you learn how to draw, then you learn how to build, and then you are an architect," I knew that I had discovered a definition of architecture that was straightforward and to the point. That disarmingly simple, but startling statement uttered by Mies represented something that upon reflection established the quintessence of architecture as I came to understand it. That statement also denoted a focused person; someone who understood his goals and had the discipline required to accomplish them.

In every encounter that I had with him, Mies van der Rohe's rock-solid persona set him aside from all other architects. Mies seemed to have his own unique access to truth, and I confess, I was awed by that quality in his persona. It suggested that his comprehension about the fundamental nature of matter was anything but superficial.

For the past four decades I have resolutely lived in one or another apartments of Mies van der Rohe's Lake Shore Drive glass apartment towers because both the buildings and their architect represent a measure of excellence to which I aspired and by which my life, as well as my work as an architect, might be continuously challenged.

*   *   *

A major influence in rounding out my perception about architecture was the enormously talented albeit eccentric loner, the wunderkind architect Paul Marvin Rudolph. As Yale University's demanding architecture school chairman (1958–65) Rudolph force-fed his students the idea of the discipline implicit in the work itself, demanding that I give credence to the importance of "willing" something into existence. In later years, when I expressed my excitement to him about one or another of my projects before it was actually completed, he stated enigmatically in Gertrude Stein-like terms, "When it's built it's built." Implicit in that statement is the uncertainty about architectural accomplishment adjudicated before it comes into being.

In other words, he wouldn't pass judgment until the construction of the building was concluded. He made me understand that causing a building to transpire was not something accomplished nonchalantly, rather it represented an attitude that required massive amounts of discipline, and in most ways that I understood was utterly distinct from imagining a building in the first place. Rudolph's principles virtually separated architecture into two distinct parts: its concept versus its construction, each one of which seemed to be no less important than the other, yet together seemed to be symbiotically interdependent for reification.

For Paul Rudolph, architecture and life were inextricably intertwined. He lived, breathed, slept, lectured, taught and of course practiced architecture. I was thoroughly bedazzled by the depth of his commitment to unpacking the never-ending layers of space and mass that architecture represented for him. He was, other than simply being a superb teacher, the consummate architect, if by that one means a person who, in a Zen-like sense "became" his work. In all these years of practicing the discipline, if there was anyone I met who "walked-the-talk" it was Paul Marvin Rudolph.

Paul Marvin Rudolph observing 1960 Bachelor Thesis Jury, Yale University; New Haven, Connecticut; ca. 1960

As I saw it, one less-than-laudable result of his single-mindedness represented a poignant one-dimensionality, an aspect of Paul's life that positioned him for me as a kind of tragic hero. Aside from architecture, his interests seemed to have neither the depth nor breadth of the magnitude of his focus on his work. Since Paul Rudolph was my teacher over a two-year period I had ample opportunity to assess his pedagogical skills and after my time at Yale, I saw him on numerous occasions leading me to consider him a mentor of some consequence. Since a number of his buildings had the inexplicable quality to which I alluded earlier, one could make the assumption that "aura" was an important factor that he might communicate as a teacher. Sadly, that was not the case, or at least in my judgment he never considered it in his critiques to any degree that I was aware of.

*   *   *

By contrast, the next force that influenced the way I felt about architecture (more a colleague then a literal mentor) was the native New Yorker John Q. Hejduk, the 6-foot 8-inch tall "gentle giant" who showed me by example that "paper architecture" was indistinct from its actualized counterpart and

Antoine Grumbach, John Hejduk, the author and Charles Moore in Paris, France, 1975

was, in his view, its equal. For example, because of its iconic image, Hejduk's concept for his well-known Wall House project conceived of and drawn in the 1960s was at least as memorable as the reality finally built in the Netherlands after his death in 2000. The mystery implicit in Hejduk's Wall House was immediately apparent in the drawings while the actualized form never added a significant dimension to that which was implicit in the design. Of course, that's one of the main projects of architecture, to retain that inexplicable quality that is sometimes diminished by its three-dimensional actualized presence.

I would go so far as to say that once a concept is clarified through drawings and models, it is as close as it is ever going to be to its actualized form. If not, then some unfortunate synapse has come to pass confounding the result; thus my revulsion towards that unattractive phrase "value engineering;" that ludicrous term whose purpose it is to dumb down a project across all levels before it fulfills its promise portended by the drawings that foretold it. Hejduk himself persuasively argued that, "the closer a finished building was to its original concept, the more poetic—and, by extension, the more authoritative it became."

His concept of his heroically-scaled "House of the Suicide" and "House of the Mother of the Suicide" architectural sculpture, now constructed in the entry hall at the architecture school at Georgia Tech, was only made more heartrending by the finished three-dimensional piece, but I understood the gravitas of the thought underpinning it was to be found in the sketches and finished drawings that preceded its sculptural actualization in Atlanta, Georgia.

Both conceptually and actually, Hejduk persuaded me that the real challenge implicit in architectural production was to embed passion into one's work. He felt strongly that architecture was "a passionate pursuit" and he uttered that phrase on too many occasions to doubt his motives. He was deeply suspicious of those designers who maintained distance from their work giving the impression that architecture was not much more than an intellectual pursuit, or at most a geometric exercise. John's poetic posture about building gave credence to the importance of art as it informed architecture, thus enriching it.

*   *   *

John and I first met in Cornell University's Student Union in the autumn of 1963. We were both teaching at separate architectural design studios in Ithaca, and more to the point Cornell had also invited me to mount an exhibition of my paintings in the architecture school. One fall evening while I was enjoying a quiet supper alone in the rabbit warren-like Gothic student union building that is Willard Straight Hall, I spied an enormous creature rapidly descending upon me. Without the nicety of an introduction, he immediately alleged argumentatively that because my paintings were organized on the diagonal they were copies of some of his architectural drawings that were similar in format. When I tried to explain that I didn't even know who he was, my confession did little to assuage his aggravation. Nonetheless, after a prolonged exhaustive vent, he calmed down long enough so that we could begin an aesthetically inclined conversation that lasted well into the night and onward for 37 years. John helped me to validate the importance of poetics that he felt were implicit in architecture.

Later that fall, Hejduk and I were on the same final architectural jury at Cornell that coincided precisely with the assassination of President John Fitzgerald Kennedy. In our case, that tragic event was immediately followed by our silently walking the Cornell quadrangle together with an equally silent tearful and stunned student body. The horror of that time did much to solidify our warm friendship that endured until Hejduk's untimely death in 2000.

Hejduk's fascination with the inexplicable in architecture was more important than any other factor in his architectural persona and had a profound impact on my perception about the discipline.

*   *   *

Last but hardly least, a self-proclaimed "man of the people," the Bangladeshi architect Muzharul Islam made it eminently clear to me throughout successive decades of endlessly circular ideological argumentation about capitalism versus Marxism that architecture was far more in the service of those most in need of it than for those who claimed it their right simply because they could afford it. His reiterative rant was something one either bought into entirely, or rejected out of hand.

For me, my friend Muzharul Islam was a paragon of ethical behavior that extended into his professional practice, no matter the self-defeating implications that such a posture might mean in any practical sense. His sometimes unnecessary outspokenness on the subject put his life into jeopardy on more than one

Muzharul Islam with the author standing in front of the Vastukalabid-designed family residence in Dhaka, Bangladesh, 1967

occasion, particularly in a part of the world where those who are in highly visible positions of leadership and who are outspoken about their beliefs are in real danger of being assassinated. Pakistani Benazir Bhutto's violent death in 2007 is just one high-profile example in the succession of those tragic statistics.

Muzharul Islam utterly refused to associate himself with those persons in his country who would make a profit at the sacrifice of doing what might otherwise be important for the sake of the poor native Bengali. His determination to volcalize his beliefs was such that his architectural practice suffered increasingly throughout his professional life, but his integrity as it related to architecture in the ethical context of social cause was never compromised.

Nonetheless, it took many years for architectural students in Bangladesh to appreciate his populist position because he was harshly critical of them should their work not elevate social issues above aesthetic, or even personal ambitions. For Muzharul Islam, the purpose of architecture was to be entirely in the service of those whose needs for basic services transcended all else. Islam's total devotion to social cause precluded any aesthetic or poetic development in his work, thereby regrettably limiting his otherwise humane approach to his work.

*   *   *

I am under no illusion that what I gleaned from these five very different, but influential architects was necessarily what they intended. Furthermore, since each one so enthusiastically believed in the virtues of his own approach to the discipline, I could never quite clarify for myself the legitimacy of one point of view over and against any other one.

If I were to evaluate those projects that I thought were autonomously derived by me, I would likely be stunned by the paucity of what I might otherwise think was original to me alone. More likely, I would be pleased that the seeds sown by these pivotal predecessors bore fruit—hybridized to be sure—but after all, Americans are nothing if not a case study in hybridization.

*   *   *

If there was anything unique in what I contributed to the field over five decades, it was to be found in the idiosyncratic way that I went about the business of addressing or contending with architectural issues as they became evident. Simply put, I didn't have a master plan in place, or an agenda of some kind, nor did I ever demonstrate much of an interest in synthesizing anything that I discovered. What I did accomplish was more in the realm of edification not just within the context of conventional architectural education but within the larger realm of coalescence.

From the point of view of practice, what follows then, is less about what I may have attempted to accomplish than it is in part about the unresolved dialectic between faith and interpretation. Put another way, what follows is also about the split between classicism and romanticism. As such, this manuscript is more about the human spirit, the indefatigable human spirit that always seems prepared to face up to whatever obstacles stand in the way of progress and to do so presumably without a prearranged direction towards an unpredictable goal. Needless to say, for every conflict that I appeared to resolve, another one appeared. Throughout much of my career in architecture, I didn't understand that perhaps expressing conflict through the medium of building might conceivably imply conflict resolution.

*   *   *

Attending public high school in Chicago, I built an unstable staircase that was composed of one academic misstep after another. I have spent virtually a lifetime of remedial self-enculturation attempting to correct a series of obstinate decisions that I made in haste. To my surprise, my prematurely misguided decision-making would unintentionally play a pivotal role in generating my life-long self-educational pursuits.

Without these achievements I would never have come to grips with certain cultural issues that I would make use of both in architectural conceptualized terms and critical writing as a dialectical means of interpreting those issues that were of importance to me. As I sought some sort of legitimacy through language of what I was beginning to engage in architecturally, it turned out that the juxtaposition of architecture and writing could in turn test the other. It is also clear to me in hindsight that language offered the possibility of the security of the home I never fully realized in architecture.

*   *   *

My problems began when I deliberately opted out of Senn High School's College Preparatory Program, replacing it with the school's technical coursework because I wantonly misinterpreted MIT's admissions requirements that forswore any foreign language requirement. By cleverly dodging whatever humanities and language courses that were available to me, I unthinkingly circumnavigated the development of rigorous study habits, which proved to be a disastrous omission later on at MIT. In truth, I felt that I could get better grades in high school because the shop and drafting courses would be easier than the college prep program, but that path of least resistance in due course led to a dead end.

Whatever minimal study habits I did acquire in high school were held hostage to my growing interest in progressive jazz. By my freshman year, my parents were able to afford a rickety second-hand upright piano, one from which sound—however tinny—actually emanated and I persuaded my mother to allow me to replace classical piano instruction with jazz lessons. I auditioned with the great Sharon Pease in his downtown studio, who limited his stable of students to a select few. In addition to his small coterie of would-be jazz musicians, he edited the piano section for *Down Beat* magazine. Over the years, his name also appeared in several noteworthy books on progressive jazz, such as Art Tatum's biography. He penned numerous articles in *Down Beat* that helped to shed light on boogie woogie in the context of jazz. Finally, he also trained several progressive jazz musical virtuosi among whom was the renowned jazz pianist Johnny Guarnieri. Needless to say, Sharon Pease was sought after in the world of progressive jazz and I was fortunate that he took me on given his limited number of protégés.

Whenever Oscar Peterson, George Shearing, Art Tatum, Teddy Wilson, and other prominent jazz pianists of the 1940s came to town, a number of we progressive jazz buffs listened in awe to their performances which were held in an assortment of scruffy, smoke-filled clubs around town, whose besotted, noisy regulars were anything but progressive jazz cognoscenti. I toyed with the notion of becoming a jazz pianist until I reached the unhappy conclusion that my aptitude for improvisation was less than ideal, which when combined with my minimal rhythmic skills, strongly suggested that I opt out of such a future occupation.

Though my interest in jazz piano didn't do much for my study habits, even during high school drafting classes I began to understand that syncopation per se could be translated architecturally with some astonishing results. Over the ensuing decades, that particular perception has been ratified by a slowly increasing understanding of how differentiated musical tempi might inform architecture, or at the very least, how music and mathematics might coincide so as to impact numerical ratios, which in turn would surely impact architectural form.

*   *   *

Nicholas Senn High School was an anomaly insofar as its reputation was upscale, yet it was populated by many inner-city students of mixed academic capabilities combined with differentiated social skills. The school was a little like my grandma's boarding house filled with diverse constituents. Gentiles didn't for the most part fraternize with Jews (Hugh Jacobsen always reminded me that his gentile fraternity did not see fit to interact with Jewish fraternities) and it seemed as if every other day after school there was a scheduled fist fight between "tough guys" and "jocks" in the large empty lot behind the school. When I graduated in 1948, *Life* magazine ran an issue on American inner-city public high schools. Senn High School was featured in that issue as an example of "one of the worst high schools in the nation."

I escaped Senn's Reserve Officer Training Corps (ROTC) program into which I had rashly enlisted in my freshman year by accepting an open call as a sophomore to try out for the high school's varsity fencing team and was fortunate enough to be selected by its coach Arnold Shaffner. Members of Senn High School's fencing team honed the skills they had acquired with Coach Shaffner by being co-coached by Alvar Hermanson, the US Olympic trainer in fencing who was in residence at the University of Chicago. Those of us who regularly plied our way via public transportation to Bartlett Gymnasium in Hyde Park noticeably enhanced our fencing capabilities.

Why fencing? Because everyone connected with the sport knew that Hungarian saber fencers had historically been finalists in their weapon in many Olympiads through to the 1956 Hungarian revolution. In that year, a number of Olympic-class saber fencers emigrated to New York City from that beleaguered country, thus establishing the high level that the sport reached both at the New York Athletic Club, as well as at Manhattan's Salle Santelli. Since I was a Hungarian "thoroughbred," I thought that my national origins might help me attain some degree of success in that sport, and up to a

Illinois state champion Senn High School's varsity fencing team with the author in center rear, 1948

certain level of proficiency, they did. A walk-on candidate initially as a sophomore, I eventually captained Senn's fencing team when we won the Illinois State Championship in my senior year. At that time, Illinois high school fencing teams were only trained in the use of the practice weapon, or foil.

I didn't switch to saber until my freshman year in college when I made MIT's freshman team once again as a walk-on candidate. Trying out for the 1956 US Olympic squad but not nearly qualifying closed another chapter on my concentrating on a field in which I wasn't apparently suited. Nonetheless, my connection to fencing has endured.

In 1961 while in graduate school, I coached Yale's varsity saber squad, traveling with them to intercollegiate meets as well as accompanying them to the NCAA tournament which was held at Princeton that year. For a time in 1995, I suited up to work out at the Chicago Salle 2000, even entering an open meet at Northwestern University where my left-handed attack garnered several touches against a much junior opponent until senior legs gave way and I hung up my sabers for the last time.

*   *   *

I remember three high school classmates in particular who, like me were enrolled in Mr. Robertson's architectural drafting classes. This bow tie-clad, tweed-suited, dapper little man was a support system for those of us who perennially needed help in developing our budding drawing skills. All three students were more talented than I, yet none of them pursued architecture as a life's work. My conviction about the level of their talent as opposed to mine was confirmed years later when I uncovered drawings I had made in that high school drafting class. My mother squirreled away all my work and when she died I rummaged through the assortment of my architectural drawings that she had saved. As a result, I can absolutely verify that the level of my work was less than dazzling.

However, something became apparent in those architectural drafting classes that encouraged me to focus more intensely on architecture as a promising career. While my natural aptitude for coming up with inspired ideas was, in my view, far from extraordinary, I discovered that I thrived on repetitious tasks that didn't require much thought; cross-hatching steel, dotting concrete into basement foundation walls, etc. Excelling at these tasks required a level of precision that appealed to me, particularly in light of my limited attention span. This anomaly didn't escape me, even as I wondered how on earth such a disjunction had come about in my otherwise scatter-shot approach to most endeavors I undertook.

While I always held down part-time jobs outside of school throughout childhood, I believe that my interests in seemingly mind-numbing tasks came about because of an exacting early high school summer job where I was employed in the Seeburg pinball machine factory. My tedious task was to drill holes into three-quarter-inch thick, 4- by 8-foot plywood sheets. The sheets were propped up vertically and I placed a template over them to locate what would later become the pinball stops. While the job helped me to build up strength, which I put to good use on the fencing team, it was physically challenging and repetitious, yet for some reason I can't put my finger on, I actually thrived on that rigorous routine.

*   *   *

My curiosity about abstract art in its most reduced form came about while attending a freshman high school art class. There I was exposed to the Dutch theologian/philosopher Dr. M.H.J. Schoenmaekers, who was the champion of the spare "lowlands" abstractionist painter, Piet Mondrian. I was fascinated by Dr. Schoenmaekers' theosophical ideas relating to dualistic concepts, such as man = vertical (life) and woman = horizontal (death). Those dialectically polar opposites were posited by Dr. Schoenmaekers who used them to philosophically elaborate upon the artist Mondrian's geometrically reductive work that employed the horizontal and vertical spatial divisions for which the artist became renowned. I assiduously studied the few Mondrian paintings that were in the collection of the Art Institute of Chicago, read a book in the Chicago Public Library about Mondrian's life and work, and I was hooked. It was to be expected that I began to draw, and eventually to paint in a way that owed a great debt to the Dutch maestro.

Beginning in high school and throughout my years in the Navy and beyond, I painted first in the manner of Mondrian and then later, when the former Bauhaus Professor Josef Albers influenced me through the vehicle of his celebrated color course that I took while I was a graduate student at Yale, I painted in this manner throughout the 1960s when I returned to Chicago to practice architecture.

A painting in oil on canvas by the author in the manner of Mondrian, 1950

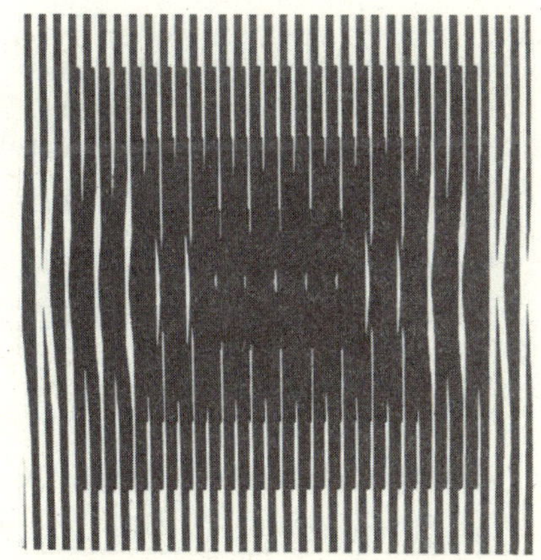

A painting in acrylic on canvas by the author in the manner of Vasarely, 1965

The work of other architects has also been influenced by their artistic endeavors. For example, Richard Meier's Marca-Relli-like collages had an impact on Richard's use of "layering" in his architecture. In my case, my fascination with geometry not only had on impact on my painting, but found its way into my architectural production as well.

The minimalism exhibited in the work of Mondrian, Albers and members of the Dutch De Stijl movement, such as Theo van Doesburg, struck a responsive chord in me. I soon found that to paint in such a way, was to subject one's self to a discipline to which I was inherently attracted. Like the detail work that I enjoyed so much in mastering the routine tasks necessary to draw immaculately, here was a similar methodology in this particular approach to art to which I could subscribe and in which I could excel.

My mother did not approve. Her observation that even she "could paint like that" didn't exactly buttress my emerging interest in geometrically reductive modernism. Unfortunately, I never took the trouble to explain to her why I was interested in this approach to art.

My painterly interests shifted from Mondrian to Albers because I could somehow never sustain an interest in developing the asymmetrical compositions intrinsic to Mondrian's work. Once again, the unbalanced characteristic endemic of modernism baffled me. On the other hand, while Albers seemingly pursued an equally abstract agenda to that of Mondrian, Albers' use of symmetry both at a macro as well as at a micro level reinforced my own natural tendencies—or comfort level, if you like—that I would continually reinforce with the buildings I designed throughout my professional life.

*   *   *

An incident occurred on the very day of my high school graduation that would have a profound effect on me. The morning of the graduation, my best friend Alan Rose and I hiked for hours on trails paralleling Lake Michigan, misjudging the time and nearly missing our own graduation ceremony. As imminent aspirants to the next stage in their life often will, we myopically ruminated simplistically about our ambitions. Our conversation rambled on about our dreams for the future and our hopes for what might transpire after high school. Alan said that he looked forward to adulthood and to the limitations that maturity would bring which might help to focus his life's ambitions. Thoroughly astonished, I countered by arguing that the promise of life was not about constraints but about unforeseen possibilities and the challenges that the unknown might bring for us. After that confrontational exchange, our closeness was never the same.

Something else occurred that spring of 1948 that was to have ramifications later in my understanding about disappointment emanating from loss. While attending the Olympic trials in track and field at Soldier Field, I watched in disbelief as the great hurdler of the day, Harrison Dillard, was disqualified in his specialty the 110-meter hurdles when he knocked over a great many hurdles, thus disrupting his rhythm. Dillard's track coach subsequently persuaded USOC officials to allow him to enter the 100-meter dash instead and he barely made the team in an event that he had never run before.

Later that summer in the 1948 London Olympics, Dillard was the gold medalist in the 100-meter dash. I didn't know it at the time but disappointment was to be my companion in no time at all and then periodically throughout my life and, like Dillard, I would need to find ways to cope with it, indeed to make something positive come out of such confounding situations.

*   *   *

Classic teenage indolence combined with willful scholastic choices severely limited the development of my study habits such that when I finally matriculated as a scholarship student at MIT in the autumn of 1948, I was utterly unprepared for the rigors of that institution's quantitatively driven educational demands. In hindsight, I appear to have done everything in my power to flunk out of college perhaps sensing intuitively that my parents could not in any event have afforded to have me continue my education at MIT.

Joining a fraternity, switching weapons from foil to saber and fencing on MIT's freshman team, becoming the coxswain of the freshman lightweight crew team, playing piano gigs after midnight on Boston's WHDH radio station and doggedly dating a Brookline high school senior on her way to Wellesley, whose mother felt that I wasn't good enough for her daughter was almost, but not quite beside the point. It seems as if I engaged in these many activities so as to assure that I did not have the time necessary for the rigorous study of subjects that were in any case beyond my capabilities, as well as requiring a level of diligence for which I seemed incapable.

The author (right) with his parents Emma and Sam at the Chez Paree nightclub in Chicago, 1949

None of these excuses disguise the fact that I couldn't handle the required freshman coursework, which included differential and integral calculus, differential equations, radio chemistry, physics and descriptive geometry. Just noting the subjects that confronted my unprepared self in the fall of 1948 sends shivers down my spine even now, six decades after the fact.

However, the one arena in which I excelled during that inauspicious year in Cambridge was putting in countless hours drawing while fantasizing about architecture. I discovered that while I didn't necessarily have talent in any conformist sense, I had the capacity to work longer hours by far than my contemporaries on things that I wanted to do. My problem was that I lacked the discipline to dedicate the necessary time to what I needed to do.

*   *   *

Flunking out of MIT was clearly a no-brainer, and when the undergraduate dean of that august institution informed me at the conclusion of my freshman year that my grades were "the lowest recorded in the most recent half-century at the school," I naively sought with great bravado to persuade myself that my dismal record was actually a badge of honor. Fifty-five years later in 2004 compelled by some kind of ironic pique, I belatedly acquired the transcript of my freshman grades. That discreditable document has since been framed and mounted and is brashly bookended by my two Yale degrees. For the next decade, I couldn't face the humiliation I felt. Only after acquiring my two degrees from Yale a decade after the fact was I able to acknowledge that less than scintillating performance in Cambridge.

Tail between legs, I retreated back to the Midwest. My parents seemed glad to have me home, although the circumstances surrounding my return were anything but heartening.

*   *   *

In spite of my unremarkable performance in my first and only year at MIT, William Wilson Wurster, the well-regarded San Francisco Bay Area architect and then dean of MIT's School of Architecture, was compassionate enough to recommend me for an apprenticeship position with his old friend and colleague, the respected Chicago residential architect George Fred Keck. While I surely didn't take the job for granted, I didn't entirely understand at the time just how fortunate I was that my initial introduction to professional practice was not only with a recognized master, but with a far-sighted architect virtually a half-century ahead of the curve on the subject of "sustainable architecture."

George Fred Keck (everyone referred to him as Fred) was something of a non-conformist not entirely unlike Ayn Rand's protagonist Howard Roark in *The Fountainhead* and for that matter, Ms. Rand's acknowledged influence in her portrayal of her idol Frank Lloyd Wright.

A loner, George Fred Keck never joined the American Institute of Architects and seldom engaged other architects discursively. He also did precious little to advance his own cause either through publishing his work or by submitting his projects for design-awards programs. Quite simply, he was his own man, a conflicted independent cuss to be sure, but nevertheless a splendidly talented architect and an individual who steadfastly resisted being either influenced or categorized by his peers. Fred Keck's younger brother Bill ran the production side of the office, but it was Fred with whom we in the drafting room identified, it was Fred who we all admired and it was Fred who we all tried to emulate. In my own case, I found it convenient to interpret George Fred Keck's intractability as an enviable characteristic that made him distinct from a rather more conventional understanding of architecture.

Aside from his abbreviated experience teaching at the Institute of Design, Fred's interaction with his office staff was situated somewhere between rudimentary and perfunctory. Since he never discussed aesthetics in the office, we assumed that this subject did not loom large on his radar screen.

Nonetheless, George Fred Keck was an exemplary role model who cut a remarkable figure with his carefully coiffed mane of salt-and-pepper hair. Stiff-backed and entirely erect, and like Paul Rudolph and Harry Weese who also served as commissioned officers in the Second World War, Keck would stride purposefully through the drafting room puffing on his briar pipe. In summer, dressed in his wrinkled white-linen suit, white shirt, black bow tie and white bucks he would make abbreviated stops at each of our drawing boards always prepared with concise, or in my case terse, "crits" (critiques).

With his natural reserve, he maintained distance between himself and his staff. I cannot recall a single example of casual conversation that he ever had with any of us. It was rare that he even removed his jacket in the office. Nonetheless, while the staff respected Fred, I was in awe of him; but it was his female patrons who adored him. As the picture-perfect image seemingly out of central casting of a mid-20th century American architect—much like the architect portrayed in the Cary Grant, Myrna Loy film *Mr. Blandings Builds his Dream House*—Fred was absolutely charming when he wanted to be, and when he personally painted a water-color perspective rendering showing the "lady of the house" the proposed design of what she could expect from her architect, she was all a-twitter.

But it was the science of his pioneering, environmentally responsive architectural production that was truly spellbinding. After Keck returned from the Second World War, he and Cambridge's Carl Koch (who coincidentally taught at MIT's School of Architecture), were among the first American architects to consistently orient their buildings on a north–south axis for purposes of maximizing the available environmental impact. This alignment responded to the importance of accommodating their micro-region's sun solstice. Working within the constraints of the local solstice, Keck always incorporated into his designs on the south façade horizontal overhangs which he cantilevered dimensionally that hovered above glass walls in such a way that the sun would penetrate deep into the house warming it in the winter, but the blazing hot sun would never find the heart of the house in the heat of a Chicago summer day.

He customarily situated narrow double-glazed strip windows above storage closets on the dwelling's north side so that in the winter the storage units could act as a physical buffer that helped to insulate the residents against Chicago's frigid north winds. Cork floors anchored with mastic onto poured concrete floor slabs were always radiantly heated with hot water coursing through copper piping making the cork toasty beneath one's toes, a far more healthful mechanical system than blowing hot air on the inhabitants—something that I discovered three decades later that the Romans had invented two millennia earlier.

Fixed, double-glazed panes of glass were always discretely separated from opaque louvered and screened ventilation hopper panels on both north and south façades. These panels always provided through ventilation long before air conditioning with its obscenely excessive energy requirements came on the scene. Characteristically, Keck avoided placing windows on the east and west façades because there was no way that he could successfully screen the rising and setting summer sun from penetrating the interiors of the house.

Whenever possible, Keck employed exposed masonry bearing cross-wall structures that were spanned by exposed heavy timber beams. These, in turn, supported dead-level roofs which were designed to hold water so as to further insulate the rooms below. His projects were invariably modernist, mostly one-storey houses that were reductively informed and painstakingly proportioned. Carefully crafted, they were bereft of any ornamental program, artifice or embellishment whatsoever and his ground-breaking structures were extremely popular with Midwestern clients in the mid-20th century.

It was all proto-environmentally eye-catching and because I had no other experience to measure against it, I naively thought that everybody designed with environmental accountability in mind. I missed entirely the ethics implicit in his innovative skills that conserved fossil fuel expenditures. The first year of anyone's apprenticeship is invariably influential, and mine was no different, although 30 years would pass before I would consciously find a way to incorporate the information I gathered in Keck's office into the design of my own buildings.

Still, because Keck's houses seemed to be entirely rational in their conception, I felt something that I couldn't quite put my finger on was amiss. I was troubled that there was nothing inexplicable about Keck's architectural production, something that I felt was necessary for it to qualify as art.

*   *   *

During the all-too-brief year that I was employed by Keck, he attempted to persuade the Chicago Housing Authority (CHA) to commission him to design a single-loaded high-rise apartment building with a south-facing exterior gallery so that he could employ his energy conscious lessons in public housing for the benefit of those most in need. Toward that end, in late 1949 Keck asked Ludwig Mies van der Rohe to support his quest even though their relationship was frankly perfunctory. He invited Mies into the office when Elizabeth Wood, then the executive director of the CHA had scheduled a visit in an attempt to convince her of his credibility. Although he succeeded in his bid to get the CHA to hire him, unfortunately, Keck's use of chain-link fencing material on the open south-facing galleries of that high-rise became an unintentional negative symbol to latter-day behaviorists who missed entirely the energy conscious project's raison d'etre. While there is little doubt that socially conscious issues were not on Fred Keck's radar screen, his ideas about ecological sustainability were nothing if not commendable, and his efforts in that regard were a full half-century ahead of the curve.

*   *   *

George Fred Keck exploited his reputation by helping to bring about the New Bauhaus, whose name was later changed to the Institute of Design, and he was instrumental in bringing Lázló Moholy-Nagy to Chicago to head the program. Just after Bertrand Goldberg returned from the Bauhaus in Berlin, he worked for Keck, who headed the architecture program at the New Bauhaus from 1937 until 1942.

Coincidentally in 1937 as well, Mies van der Rohe came to Chicago to head the Armour Institute's architecture program. Fred Keck's feelings were hurt by Mies's maneuver in the early 1940s to displace him from the Institute of Design faculty. This ploy was part of Mies' attempt to consolidate architectural training solely within his architecture department at the Illinois Institute of Technology, thereby voiding it from the ID's curriculum. In all events it was common knowledge that Mies and Moholy-Nagy had a tense relationship exacerbated after Moholy-Nagy's death by his widow Sybil. Even so, Fred managed to put all of it aside by inviting Mies to his office. The gesture never accomplished much in the way of healing their rift, although they were always civil towards each other whenever their lunchtime coincided at Chicago's Arts Club.

At Yale a decade later, I was present to hear a lecture given by Moholy-Nagy's widow Sybil when she castigated Mies for adding his name to a letter circulated by the Deutscher Werkbund supporting Adolph Hitler in 1933. Apparently, deep-seated grudges die hard.

Being present at such a meeting held in Keck's drafting room about the value of environmentally responsible architecture in the context of social cause made an enduring impression on a 19-year-old wannabe architect who had just flunked out of MIT. Even so, a generation would pass before that naïve young man fully understood the value of the conjunction between social and environmental accountability.

My societal "greening" evolved further a few years later when my friend Alan Rose's father Alvin M. Rose succeeded Elizabeth Wood as the executive director of the CHA. Before Mr. Rose's services were peremptorily terminated by the notoriously controversial CHA Chairman Charles Swibel, Mr. Rose and I spent time on Chicago's skid row witnessing incognito cultural patterns common to many who were to become residents of the CHA.

*   *   *

It was blind luck that my apprenticeship was served with an architect of such distinctive vision combined with a unique aesthetic sensibility. And apprenticeship is what it was. I was paid $15.00 a week for the first six months, and was thrilled to get a raise to $17.50 for the next six months. Fred Keck gave me the earliest, and one of the most priceless pieces of advice I ever got from an employer. He told me to learn as much as I could about the making of real buildings over all else; particularly design.

Keck said, "Do working drawings, check shop drawings, and get field experience until you cannot stand it any longer … until you are ready to throw up rather than do any more working drawings. Leave design for later, much later." I understood his advice in the context of "materiality" versus "spatiality" and my work ever since has been far more developed in terms of what a building is made of versus the spatial sense that more often than not motivates other architects. Not knowing any better, I obediently took his "nuts-and-bolts" advice and for nearly a decade, I did just that.

*   *   *

When he was there, the gaunt Robert Bruce Tague perched on his stool at the drawing board behind me at the north end of the small, yet sunlit sixth-floor drafting room. I credit this eccentrically talented, yet self-defeating chief draftsman with teaching me architectural nuance and good taste—connoisseurship.

Robert Bruce Tague's Frueh House, Highland Park, Illinois, 1949

Needless to say, Tague's role in the office was more far-reaching than that. As Keck segued out of his 1930s International style depression-modern design phase into environmentally responsible architecture, Bob Tague remained the conscience of International style modernism in the office. Nonetheless, Tague and Keck benefited from a symbiotic association in which each influenced the other to whatever degree that was possible, given their diametrically opposed personalities.

Robert Bruce Tague, 1949

It was well known at the time that Bob Tague's own professional career could have become far more distinguished than it did but for his penchant for alcohol. Combined with his instinctively soft-spoken, extremely shy persona this obsession distracted him from ever achieving any great visibility. Like some other Chicago architects of the period, Tague was periodically hung out to dry at Riccardo's, the Art Deco Rush Street watering hole two blocks from the office that was frequented by architects, artists and designers many of whom were on the faculty of the Institute of Design.

After Tague's military service as an enlisted man in the army in the Second World War, he stayed on for a time in Europe, becoming a part of a collection of avant-garde artists and architects in Paris that included the already-famous Le Corbusier, the presciently well-known art gallery dealer Denise René and, among others, the artists Hans Arp and Sophie Tauber Arp. It took me 27 years to acknowledge my debt to Tague, but in 1975 it gave me a great deal of pleasure to include an example of Bob Tague's domestic architecture in the revisionist exhibition *Chicago Architecture* that I helped to curate. Swallow Press published the catalogue of the show.

Day in and day out, during the period that I apprenticed to Keck, Bob Tague incrementally advanced my knowledge of architecture all within the lexicon of translating Fred Keck's freehand design sketches into built form. Decades before the use of computerized drawings replaced drafting by hand, all of Keck's draftsmen drew beautifully. In fact, one of Keck's most accomplished draftsmen was the now-deceased, former University of Minnesota Architecture School Dean, Ralph Rapson. Encumbered by having the use of only one hand and using his eye only as a measuring tool, he could hatch in pencil perfectly closely spaced lines depicting vertical wood siding or a cant strip on elevations of Keck's imagination.

But none of them was a more gifted draftsman than Robert Bruce Tague himself. By emulating his motor-control skills and exerting an entirely unheard amount of patience, I found that, with many long hours of assiduous practice that reinforced my inherently compulsive nature, I could take pride in accomplishing those same ends.

I learned to prepare immaculate presentation drawings using an old-fashioned India ink pen with nibs as well as to twirl a perpetually sharpened 4H mechanical lead pencil as the point met the vellum so as to produce consistently dimensioned pencil lines for the few working drawing assignments that came my way. Little did I know how useful these compulsions of mine would be until a decade later when I was wearing white gloves as I prepared color plates for Josef Albers's famous "color course" in which I had enrolled as a part of my course work while in graduate school at Yale.

*   *   *

After hours I became Tague's night school student at P.B. Wight's stone-clad neo-Romanesque Institute of Design (ID), which was just four short blocks away. Three years after the New Bauhaus founder László Moholy-Nagy died in November of 1946 the school that this former Bauhaus teacher initiated was still on fire with revolutionary ideas about art-and-design education as well as environmental design.

During the year that I studied at the ID's night school, visionary architects and designers such as Serge Chermayeff, Buckminster Fuller and Konrad Wachsmann could be seen roaming the corridors of the ID. The haughty White Russian, British-educated Chermayeff was a confrontational type and as such we soon developed an adversarial relationship. The year before I studied there, Fuller developed his geodesic structures at the ID and had taken those students and faculty who had worked on the project to Black Mountain College that summer to erect examples of that work. During the 1949–50 academic year Fuller was back and forth between the ID and the school at Black Mountain in North Carolina.

The gifted furniture designers Bartolucci and Waldheim, originators of the unexpectedly comfortable and reductively beautiful Barwa chair (my first modern design purchase) coexisted happily in design labs at the ID with the brilliant textile designer Angelo Testa, whose geometrically patterned fabrics were specified by modernist architects, designers and decorators alike. Another important figure at the school was the sought-after lighting designer Robert Lederer whose free-formed Alvar Aalto-like, edge-lit Lucite table lamps were all the rage in the Well of the Sea cocktail lounge on the lower level of the old Sherman House Hotel across from City Hall. Artists such as John Whalley and Richard Koppe, both of whom taught alongside me at the University of Illinois at Chicago a decade and a half later, were influential in promulgating Moholy-Nagy's pioneering ideas about design and design education.

Returning from the Eliel Saarinen-designed Cranbrook Academy of Art in Bloomfield Hills, Michigan—after his naval service in the Second World War and a stint at Skidmore, Owings & Merrill—Harry Weese had a small office in the same North Michigan Avenue building as the brothers Keck. He was joined there later by John van der Meulin, who also periodically collaborated on projects with Ralph Rapson.

At the corner of Michigan Avenue and Ohio Street in Holabird and Root's Art-Deco-styled Dana Court Building, Weese's wife Kitty Baldwin together with Jody Kingrey opened the Baldwin-Kingrey furniture store that purveyed modernist furniture and accessories. The sought-after interior designer Marianne Willisch, who from time to time designed interiors for both Keck and Weese, frequented their offices on a regular basis. Edward Dupaquier Dart, who had recently returned from the Yale School of Architecture after serving as a marine fighter pilot in the Second World War, opened a small studio a few blocks away.

With such a coterie of design talent in place, the Chicago architectural scene was a hothouse as the decade of the 1940s drew to a close. Even as a novice, I thoroughly understood the importance of what was sizzling around me and I was thrilled to be an infinitesimally small part of it.

*   *   *

I think I always knew that I would never return to architecture school at MIT, and after my apprenticeship year in Keck's office, two of his draftsmen and I prematurely opened our own shop. We called our fledgling partnership Tigerman, Rudolph and Young, and maintained an uncertain practice for six months from June through December of 1950 even as the Korean War was escalating.

Clarence Rudolph was an older, very accomplished professional albeit with no particular academic training of which I was aware and Michael K. Young, our resident "registered architect" who had recently graduated from IIT was by definition Miesian to the core. My two confreres kept their day jobs in Keck's office, while I held the fort in our sparsely outfitted miniscule digs. They both came in at night to assist me in the preparation of what few working drawings Mike Young had contracted for us to churn out.

Nothing much came of our efforts, with the promising exception of a seminal meeting that I arranged with Mies van der Rohe's developer patron, the former rabbinic student Herbert S. Greenwald. Coming to that meeting in their small South Michigan Avenue office in early 1950, I presented Mr. Greenwald and his colleague Robert McCormick with some sketches that I had made of a low-rise housing proposal, naively hoping that he might hire our unassuming start-up practice for a building type I frankly had precious little idea how to effectuate.

Astonishing as it might seem in hindsight, Herb took an incomprehensible interest in me, eventually promising that I would one day design his "small" buildings, while Mies van der Rohe would continue to be the architect for his "tall" buildings. Later, he even went so far as to comment on a draft of a rudimentary text for an unpublished manuscript on architecture (*Architecture on Sand*, circa 1953-54) on which I had labored while I was in the Navy.

*   *   *

My first exposure to Mies van der Rohe and the workings of his Chicago office at 230 East Ohio Street came about because of my friendship with Herb Greenwald. I discovered there that the understanding of construction held by Mies' employees did not come solely from books. Instead, it was a product of information they gained by preparing full-size models of selective fabrication details in their very ample office, which had high ceilings that could accommodate large mockups of sections of buildings, the proportions and construction of which they were studying.

For example, when Mies's draftsmen sized rectangular steel bar stock window stops, they knew unerringly how many screws per linear unit of measure it took to mechanically anchor those stops to a substrate without warping the steel stop. Their knowledge was arrived at independently from industry standards. Mies van der Rohe's staff all seemed to have the necessary knowledge about the modulus of elasticity of that particular cold-rolled steel bar stock. Thus, I came to understand that the science of architecture, in so many words, was not just a result of product information contained in catalogs, or the word of manufacturers, suppliers or contractors, but significantly, of independent analyses by architects as well. In other words, Mies van der Rohe's staff was always prepared to design something themselves as opposed to specifying unsatisfying self-similar items picked from catalogs, particularly if the off-the-shelf product didn't precisely satisfy the particular aesthetic that best represented Mies van der Rohe's intentions.

The staff had also gained knowledge about the nature of materials through experience in the field while on construction-site inspections. These insights observed in and out of the great architect's office had an important impact on my understanding of the value of architectural practice as discrete from architectural education, and solidified the advice given me by George Fred Keck.

*   *   *

I was given my initial learning opportunity about the nature of materials in the field one frosty Saturday morning in the late fall of 1950 when Herb Greenwald invited me to observe how Mies and his Hungarian-born structural engineer, the celebrated Frank Kornacker, might contend with an unforeseen structural problem that had arisen on Chicago's near north-side construction site of Mies' first pair of ground-breaking glass apartment towers at 860–880 Lake Shore Drive.

The two steel channels laterally defining the edges of the 40-foot-long covered walkway between the two glass towers had deflected more than had been anticipated. This structural failing was due to steel production not being back up to speed so soon after the Second World War. This resulted in heavier but dimensionally comparable-sized members being delivered to the site rather than those that had been specified by the structural engineer based on information provided by the American Institute of Steel Construction (AISC) Handbook. The ensuing deflection was noticeable to the eye and required some kind of resolution that no one seemed to be able to define. Enter the structural magician Frank Kornacker.

Mies and Kornacker instructed the attending contractor's carpenters to quickly erect a scaffold running the 40-foot length of the steel members. Kornacker then climbed up the scaffold and ran back and forth along the top of the plywood covered scaffold with an acetylene blow torch heating up the top flange of the steel channels thereby drawing them up slightly as the steel expanded toward the heat. Then, before the steel cooled, he without delay doused its length with buckets of cold water to fix the channels into a permanent slightly bowed upward camber imperceptible to the naked eye. I had just witnessed a mesmerizing process and began to understand that there were hands-on aspects of architectural problem solving not always apparent in texts.

860–880 Lake Shore Drive Towers, Chicago, Illinois, 1950

I think I identified with the eccentric behavior that Kornacker had shown in his unique understanding of the problem.

*   *   *

Herb's statement that we would work together in the future turned out to be true. Eighteen years would elapse before his prophecy would come to pass, and then it would be with his successor firm Metropolitan Structures with an invitation to work on their vast housing project on Nun's Island located in the La Prairie Basin offshore from Verdun, within view of downtown Montreal, Quebec. As Herb Greenwald had foreseen, Mies's office was named the architect for the towers, and I was commissioned to design the townhouses.

Before the immense Nun's Island housing project began, however, there was an awkward but oddly amusing moment when Herb's successor, the nuclear physicist turned lawyer who subsequently became a developer, Bernard (Barney) Weissbourd, attempted to convince Mies to join the project team. A gathering was called in Mies' large drafting room with the developer and his "value-engineering" driven minions, landscape planners, architects and engineers. At such a meeting, it would be customary behavior for an architect to jump through hoops in order to plead with the client to move forward on a particular project.

This was a mostly memorable moment that I witnessed, however, for it was the client, Barney Weissbourd, who was presenting slides of Nun's Island so as to persuade Mies to join the team. However, the meeting was made almost unbearably tense as the great architect sat stoically and wordlessly in his wheelchair, cigar in hand, throughout the slide presentation.

At one point, Barney showed a particular image that had been prepared by his landscape consultants that was included to illustrate a series of curvilinear roads appealingly coursing throughout the island. The intention was obviously to produce a *gemütlich*-like ambience that they must have thought would be a nice counterpoint to Mies's rather more astringently reductive towers. Affecting a perplexed posture while observing the curving road system, Mies turned to his assistant, Joe Fujikawa and, in an accented stage whisper, observed dryly: "Joe, I did not think that the island was so hilly." Joe Fujikawa's measured response was: "No, Mies. It's flat." As I recall the discomfiture of the moment, no laughter followed that ironic exchange, just barely undemonstrative, yet inferentially loaded silence.

In any case, the gathering in Mies's office was anxious from the get go. Barney was trying to persuade Mies how handsome the architect's Baltimore office building, One Charles Center, looked when it was completed. It seems that Mies and Barney had done the project together after Herb Greenwald's untimely death, and in the spirit of what is now referred to as "value engineering," Barney's always cost-conscious development group had persuaded Mies to clad the building in hard coat aluminum, whereas Mies had originally proposed either bronze or stainless steel. Reminiscing that earlier he had had the same manner of discussion with Joseph Bronfman while designing the Seagram Tower on Park Avenue in New York City, Mies quietly told Weissbourd that Bronfman had selected bronze for the façade of the building over alternative, less costly materials. According to Mies, "Bronfman got something for his money."

*   *   *

In June of 1966, during one of my many meetings in Mies' office with his associates about master planning issues relating to Nun's Island, I wandered into Mies' own private sanctuary to study more closely his model of the Neue Nationalgalerie project in Berlin. The Plexiglas-protected model sat atop a pedestal in one corner of the room. Noticing that the June issue of Arts and Architecture was lying on his otherwise empty wood table top, I could not help observing that the magazine was open to a two-page spread about Instant City my own futurist mixed-use, mega-structure proposal.

At that moment, without warning, Mies rolled into the office in his wheelchair—after he broke his hip, he was confined to that device for the last several years of his life. Seeing that I was gazing at the magazine, he referred to my design by saying: "Ja … that is a very simple project." Given the high regard in which I held Mies, I was at once embarrassed, and yet intensely pleased by what I construed to be an extraordinary tribute from someone who was arguably the most dominant architect of the 20th century.

*   *   *

Mies van der Rohe always observed customary Prussian standards of decorum. On June 12, 1966, he was awarded the gold medal from the Chicago AIA at a black-tie dinner at the Arts Club of Chicago. The medal was conferred by my future father-in-law Paul McCurry who was president of the AIA's Chicago Chapter at the time. Paul had invited his daughter Margaret, herself an employee at SOM, to the event. At the time I didn't realize that 13 years later I would be married to Margaret. Mies, seated in his wheelchair at the top of the graceful, white-painted steel staircase he had designed, greeted all 200 guests. His former IIT student, the SOM general partner Bill Dunlap who was emcee of the event, stood next to him. When I introduced my wife Judy to Mies, he struggled painfully to his feet because it was an appropriate gesture to rise when greeting a woman.

I learned many things from Mies van der Rohe that were to influence the way I approached architectural practice. Once while I was in his business manager Henry Stoesser's back office, the great man himself came into the tiny office followed by a collection of black-suited German businessmen representing the Krupp organization. Apparently, they were interested in retaining Mies's services for a new corporate headquarters. To ascertain his availability, Mies reviewed a simple bar chart mounted on Stoesser's wall responding that, "he would not be available to work on the project for three years." His comment represented his understanding of his staff's workload.

Mies had a 20-person office and his comment explicitly noted his reluctance to increase his staff for a single project no matter how important. This was a wholly different work ethic than the average American "corporate" architectural firm that hired and fired at will, based on its workload. I was stunned by Mies's assertion that his own requirements for practice differed from the norm. From that point forward I too chose not to be unduly influenced by clients and their often willful schedules and unreasonable requests.

Another incident that impressed me occurred in Mies's office when I overheard a conversation between the architect and one of his draftsmen. The younger man was questioning Mies about a particular steel angle-roof coping detail, wondering how Mies wanted to handle the attachment. Should the screws be on top of the coping or on the front? Mies said he would think about the detail. When I was next in their office some three weeks later, I heard him telling the young man that he preferred the screws on top of the coping. In other words, Mies chose to "do extreme diligence" and to not permit decisions to be made precipitately under pressure.

Philosophically, Mies was his own man and always had a handle on the big picture. The rush to judgment that governed the lives of so many others was simply not a part of his persona. Having the opportunity to observe the way that Mies went about that particular architectural detail was to say the least, a meaningful experience for a young architect trying to come to grips with the often jarring juxtaposition between design philosophies and the exigencies of practice.

*   *   *

The decade ended sadly for Mies, and for me, with the untimely death of 43-year-old Herbert Greenwald in early 1959. Herb was on his way to a meeting in New York when his commercial flight crashed into the East River just short of LaGuardia's runway. The pilot had attempted to land in a fog and miscalculated where the runway ended.

The subsequent memorial service took place in Anche Ehmet (Herb's conservative Chicago synagogue) and was attended by virtually every architect of note in Chicago. Of the three speakers, Mies's comments were the most poignant and extremely personal. We all witnessed a moist-eyed Mies van der Rohe haltingly refer to "the loss of my great friend and supporter, Herb Greenwald." Herb may have been three decades younger than Mies, but he was not only very deferential to him but very protective as well, even going so far as to purchase an Oldsmobile Sedan for Mies that he sometimes used for vacations having one or another of his staff drive him to places like the Blue Ridge Mountains in Tennessee.

*   *   *

To return to 1950, the war in Korea was heating up, and on December 10 of that year my draft notice arrived in the mail. Fearful of being sent to the frozen trenches of Korea, my old friend from kindergarten days, Alan Rose, and I enlisted post-haste in the United States Navy. Before the year ended we had been sent to boot camp at the Great Lakes Naval Training Station, 40 miles north of Chicago.

In boot camp, I soon learned that military discipline is not without a sardonic, albeit sometimes cruel sense of humor. One bitterly cold winter night, while all 126 recruits were abed in our company's barracks listening after hours to the radio broadcast of one of the historic Tony Zale versus Rocky Graziano middleweight championship fights, the officer of the deck (O.D.) burst in ordering "lights out!" An unidentified voice in the back of the barracks yelled out: "give me liberty or give me death!" "Who said that?" barked the O.D., and the same voice laughingly shot back: "Patrick Henry, you dumb son-of-a-bitch!"

Immediately the O.D., a tough chief petty officer, had the entire company dress and quick march in the snow all that long, frigid night. The punishment doled out for the culprit, a sinewy street kid who was somewhat on the small side, was to be enrolled in the heavyweight division of the Navy's boxing version of Golden Gloves. For the rest of his time in boot camp and on a regular basis, that scrawny recruit was beaten senseless by much larger opponents.

Shortly thereafter, I came down with a case of measles and when the corpsman petty officer on duty at sick bay was told that I was allergic to penicillin, he injected me with it anyway. Lapsing into a coma with a very high fever, I nearly expired and remained in hospital sufficiently long to miss my company's graduation.

At the completion of boot camp with my new company, a lottery was conducted which alphabetically assigned the first 100 or so of our company's recruits to the Seventh Fleet in Korean waters. A few of us at the end of the alphabet were assigned to duty at the Boca Chica Naval Air Station not far from Key West, Florida.

That was the beginning of a four-year tour of duty in the military that, more than anything else, gave me the time and the support to grow up. Even so, the day that I arrived at Boca Chica, I received my own personal version of paradoxical Navy justice. That Sunday afternoon while exploring the base, I discovered a wonderful-sounding organ located in the Navy chapel. Impulsively, I sat down and instinctively began playing progressive jazz riffs. Upon hearing unexpectedly syncopated sounds emanating from the chapel, within minutes the shore patrol arrived. I was arrested, interred briefly in the brig and was sentenced to a captain's mast (the Navy version of disciplinary action shy of a summary court martial).

When my air squadron's Commanding Officer Edward Hessel discovered that I was Jewish, the punishment that he meted out was to assign me to sing in the Navy choir every Sunday at that same Episcopal Church that housed the tell-tale organ. Ever magnanimous, Commander Hessel did not require me to take communion, but I knew that I was an outsider all the same. Coincidentally, that chapel was attended by President Truman whenever he visited his winter home in Key West.

*   *   *

After hours I played jazz piano at night in a variety of burlesque clubs up and down Key West's Duval Street interspersed with gigs at a rundown roadhouse frequented by Navy personnel on Stock Island. Friends of mine, southern cracker tough guys from the Boca Chica gas pool, regularly hung out at that funky roadhouse and would often sit at the bar drinking while I was playing piano as they indulged in their own distinctive version of "odd man out." Flipping a coin, the loser was obligated to stand at the door to the establishment adjacent to the highway punching out the next person that walked in. High ranking officers, nurses, et al became subjected to their perverted sense of fun and they in turn were court-martialed on a regular basis. One inebriated evening while laughing uproariously at their antics, I fell off the piano stool breaking my wrist, thereby eliciting my second captain's mast even before I was admitted to the Boca Chica Naval Hospital.

While I was stationed at Boca Chica, an incident occurred that solidified my earlier decision to become an architect. It happened early on while I was billeted at the air station. As an enlisted man in Navy Air, I was bitten by the flying bug and flew as many "hops" as I could persuade pilots from my squadron to take me on. Our flights utilized the Navy's first two-seated fighter jet aircraft, the T2V (the Navy's version of the F-80). I flew mostly with LTJG Frank Bardecki, who was the squadron's "hot dog" aviator. He loved to buzz Cuban passenger planes, flying close enough to see the whites of the traveler's eyes as they sat in terror at his antics.

I was with him on one of his touch-and-go exercises flying between Boca Chica and an aircraft carrier anchored off Key West. Stored in the back pocket of my dungarees was a letter from my Wellesley girl friend that she (always) penned in blue ink on blue paper. As a result of panicky perspiration from the g-forces on the tight turns we made, the ink on the letter disappeared. It was when Bardecki let me briefly take the controls of the jet aircraft on a rapid ascent while telling me to look up into the fathomless blue black of the sky above us, that I was totally seduced by the concept of flying.

It was only natural then that I apply for flight training at Pensacola, Florida. I successfully completed the preliminary physical qualifications. However, although my eyes tested out at 20/20, I had a slight astigmatism that was marginally higher than Navy regulations allowed. Navy doctors were often known to look the other way on that particular matter, but as (bad) luck would have it, during my optical examination, one of the jet pilots in the squadron crashed his jet while attempting to land on the tarmac at the Naval Air Station. My examining physician temporarily halted my eye test to go with the ambulance to minister to the needs of the pilot but to no avail—the aviator perished. The doctor returned to finish my examination, stating peremptorily that, "I better be right on the money." The result was my being rejected from the Naval Aviation Cadet Training Program (NAVCAD).

If I had qualified, I would have stayed in the Navy for 25 years as a pilot thereby saving the field of architecture a great deal of aggravation. As it was, many of the pilots that I flew with while stationed in the Keys didn't survive; they were either killed in combat in Korea, through accidents on shore or on carrier duty forced to ditch their aircraft, or on leave while they were trying to make an extra buck crop-dusting at low altitudes in ancient bi-planes that were caught in down-drafts and crashed in farmer's fields.

*   *   *

As an alternative to flight school, I applied for the SeaBees (a contraction of Construction Battalions, ergo CBs or SeaBees) and was accepted into their program. After completing the course at the SeaBees school at Port Hueneme, California, I in due course became a second-class petty officer rated draftsman and was billeted at the Naval Station at Norfolk, Virginia for the remainder of my tour of duty as the art director for the Atlantic Fleet's magazine, the *General Information Bulletin* (*GIB*).

During my two-year stay in Norfolk, I acquired a part-time drafting job at night in the office of T. David Fitz-Gibbon, a conservative, talented gentleman architect of the old school. His classically-inclined practice focused on institutional work thereby providing me with a strong sense of discipline about a doctrine other

The author at his drawing board at Norfolk, Virginia, 1953

than the one that I had become familiar with while working in Chicago for George Fred Keck.

While Mr. Fitz-Gibbon's practice was wholly different than Keck's, it was just as rigorous and equally valid. Both architects shared a seriousness of purpose and a commitment to their understanding of the art of architecture. Each of them drew beautifully and neither was particularly interested in the business aspects of architectural practice, but both were clearly aesthetically inclined and both practiced what they preached.

None of this was lost on me. Both Keck and Fitz-Gibbon taught me the concept of value as it relates to architecture, never mind the aesthetically stylistic vagaries of value. I began to understand that it was important to "stay the course."

*   *   *

In mid-1954, as my military tour of duty began to wind down, I was given temporary additional duty as a combat artist and was sent to Italy on the aircraft carrier FDR to turn out drawings that would record the ceremonies at the Vatican connected with the canonization of Pope Pius X for publication in the *GIB* back at Norfolk. Our first stop was at Brindisi, on the heel of the Italian peninsula. While the carrier remained anchored offshore, I was assigned short-term shore patrol duty as part of a two-man team.

My memory of going ashore after our setting anchor is indistinct, since my colleague and I had scarcely set foot on land when an angry anti-American, neo-Marxist crowd at a political rally demonstrating on the waterfront spotted two American sailors and without delay rendered both of them senseless. I don't remember much about the incident other than regaining consciousness back aboard ship in sick bay with a mild brain concussion and a sliced upper lip. I still carry the scars of that unique Italian reception.

*   *   *

Returning to Chicago after mustering out of the Navy, and after I recovered from the shock of discovering that my mother had inexplicably thrown away all of my jazz sheet music with editing and notations by Sharon Pease, I straight away began a job search. My first appointment for an interview was at IIT with Ludwig

Hilbersheimer, the celebrated city planner. Mies van der Rohe had taught with him at the Bauhaus in the early 1930s when it was located in Berlin; four years after the school closed its doors in 1933 Mies encouraged Hilbersheimer, as well as Walter Peterhans, to join him in Chicago.

After Professor Hilbersheimer looked at my portfolio of drawings there ensued a discussion between us about an opportunity that might be available for me to teach a freshman Visual Fundamentals course as Professor Peterhans' assistant in the architecture school. Professor Hilbersheimer had some questions about why I had left MIT so precipitously after my freshman year, and without knowing that he had the transcript of my dismal grades in front of him, I lied by saying that the dean wanted me to gain experience before returning to school. Hilbersheimer, it transpired, had caught me in a lie. I was so mortified when he confronted me with it that I have never knowingly lied since, aside from nontoxic white lies told as much to amuse as anything else.

I promptly found employment as an entry level architectural draftsman in the office of Anthony Joseph Del Bianco whose registration seal in 1954 alone was stamped on drawings for over 2000 suburban tract houses. Each of the six supercharged employees working in the second-floor office of the building that Del Bianco owned at the time could churn out an abbreviated set of working drawings in just two days for those "ranchburgers du jour." Granted those contract documents were made up of just two sheets (albeit jam packed with information) and were the domestic equivalent of "scope set" office building working drawings that I mastered later at Skidmore, Owings & Merrill. Del Bianco's office was a kind of sweatshop and I absolutely adored working there under those pressure-laden conditions.

Here was a chance to learn a great deal about house construction while laboring in an environment much like a factory doing the repetitious tasks on which I flourished and hoped to master. In the sociable but competitive spirit that permeated the office, I attempted to be speedier than anyone else in the drafting room dotting in concrete, hatching steel, and generally carrying out all the little tasks that others found tedious but that on which I thrived. My obsessive behavior had finally found an outlet of sorts in A.J. Del Bianco's drafting room.

More often than not, I volunteered to work overtime without compensation so I could learn as much as possible about wood-frame house construction. Del Bianco's drafting room crew was a hard-drinking lot. Most nights, overtime work occurred only after a prolonged martini-guzzling cocktail hour at a skuzzy gin mill a block away from the office. One night, returning to the studio half in the bag, I came across an undersized garter snake slithering across an empty lot adjoining the office. Gathering it up, and twirling it in my fingers as I weaved my way back to the office, I deposited it in the office secretary's desk drawer. The next morning we knew that Bunny Busta had found it by the blood-curdling scream that permeated the drafting room. Such were the delights of working for Del Bianco.

Regrettably, in the rush to produce I made mistakes on the working drawings of some tract housing projects. My most egregious error was unfortunately reflected in a suburban subdivision composed of two-story, free-standing houses situated cheek by jowl. When I visited that particular construction site one afternoon after work, I was appalled to find that I had misdimensioned outside back stairs and stoops, mistakenly positioning them under the kitchen window rather than at the back door where they were meant to be. A number of these wrongly located stairs and stoops had been framed up before anyone noticed that they were incorrect. In those years, production housing was such that projects were built without the burden of much inspection. My boss graciously forgave my boo-boo and paid to correct it out of his own pocket. I never forgot the incident or his generosity.

On the surface, A.J. Del Bianco appeared to be a run-of-the-mill builder/architect. He was married to Marge Valenti, the daughter of Joe Valenti a prominent Chicago home builder who provided a great deal of work for the office. Beyond such exigencies, A.J. was not without a great eye. He could saunter by one's drafting board scarcely scanning what was being drawn, quietly muttering as if to himself: "2-foot 7-inches." His comment referred to a 2-foot 6-inch door misdrawn at one-quarter-inch equal to 1 foot. A miniscule mistake to be sure, but Del Bianco's perception of such minutia was, to my way of thinking, astonishing.

He could also render freehand like an angel. His soft 2B pencil small-scale perspective sketches were among the most beautiful of that variety that I have ever seen. What I didn't realize at the time was that in the mid-1930s in his last year of architecture school at the University of Illinois in Urbana-Champaign (most of the employees in the office during my time there were also products of UIUC), he was first runner-up for that year's prestigious Paris Prize, finishing ahead of the great Eero Saarinen, who was himself in his final year at Yale's School of Architecture. In an elite interpretation of architecture, Del Bianco was without question an *auslander* but his human quality more than made up for his lack of ambition to go farther than he did.

Working day and night, I learned more about residential "stick-built" platform frame construction in the two years that I spent in Del Bianco's office than I ever did before or since. He may not have been a "star-architect," but his generosity in sharing planning principles and rational wood-frame construction methods with his employees was admirable. As a result, he engendered great loyalty among his staff, many of whom spent their entire careers with him, even taking over the firm on his retirement.

*   *   *

Not all of my employment history yielded such dividends. Two years out of the Navy, in 1956 and on the verge of marriage, I mistakenly left Del Bianco's office to acquire a substantially higher paying position with an architect whose work production was, as best as I could ascertain, measured monetarily rather than either ethically or aesthetically.

For example, while I was at work on the contract documents for a downtown Chicago high-rise hotel, I was instructed to achieve closure between the new structure I was working on and the neighboring existing office tower simply by employing flimsy 1-inch by 2-inch wood furring strips ramset into the masonry demising wall of the existing building. I was appalled to think that if that office tower were ever demolished, the hotel would be without cladding on a vast section of one of its façades, to say nothing about the ramifications that differential settlement between the two buildings might mean.

The architect's office was located in the back room of his plumbing contractor father's shop on the Near West Side of downtown Chicago. I lasted there a scant year when an opportunity arose to work as a junior designer in the Chicago office of Skidmore, Owings & Merrill, albeit at a considerably reduced salary, on the mammoth-scaled United States Air Force Academy project. Since I had recently married, a family council was called in the living room of our small flat consisting of my wife, her mother, my parents and me. The majority of those present felt that I should remain at my higher paying position because I now had "responsibilities".

In what can only be described as a willful epiphany arrived at, and not without some conflict, I responded to the family gathering in a way that would fundamentally alter my life as an architect from that time forward. In the spirit that "you are what you eat," I told the assembled body that I felt I would eventually be measured by those with whom I associated. Should I remain where I was, I would be so judged and people would seek me out in that context to ask me to engage in what I considered less-than-admirable work. On the other hand, I pointed out that if I worked for a firm more qualitatively driven, there was a reasonable chance that I might be sought out accordingly.

Thus it was that with great confidence in the spring of 1957 I joined the large design staff of the Chicago office of SOM. I looked forward to designing at last. In hindsight, I don't know what prompted me to make such a decision short of a strong desire to assist in the design of significant buildings, but it was a moment like none other and it had important implications for the rest of my professional life.

*   *   *

By the fall of 1955 I had decided to sit for the architectural registration examination. At that time one could qualify to take the four-day test with a combination of three years of study/apprenticeship. Since I didn't have an architecture degree, as you would expect I thought it was crucial that I become a registered architect. Not having completed any technical coursework I thought that it would be sensible to take the refresher course administered by the State. When I learned that Frank Kornacker, Mies's brilliant Hungarian structural engineer whom I had seen in action some years earlier, was teaching the structural portion of the refresher course, it turned out to be even more enlightening than the technical information that it imparted.

For the majority of those mostly young white males who took the refresher course with me, the opening session held at the University of Illinois' Chicago campus at Navy Pier on Chicago's lakefront must have seemed like old home week, or more to the point, like a fraternity reunion. Most of those in attendance were graduates of the University of Illinois at Urbana-Champaign and many of them apparently hadn't seen each other since graduation from architecture school. As they cheerfully swapped hail-fellow-well-met greetings and exchanged information about their apprenticeships, one of them smugly announced that he was a junior designer in one of Chicago's larger architectural firms and another one alleged that he was designing one building type or another for one of Chicago's more prominent architects. So soon out of school, and yet many of them seemed to be "designing."

One of them turned to me and asked me what I did. When I told him that I was a draftsman working for a developer-driven architect whose office was on Chicago's west side and admitted that I had flunked out of architecture school some years earlier, he and his confreres promptly turned their collective backs on me and resumed their collegially driven conversation.

I registered their slight and got down to work, studying conscientiously only to fail the entire nine-part architecture examination with the exception of the section on the history of architecture. The following year, and without any preparation at all, I experienced the deep satisfaction of passing the outstanding portions of the test, thus becoming registered by the State of Illinois to practice architecture.

*   *   *

Eighteen months later I was employed at SOM. From the get-go it was, to say the least, a grind, but a welcome grind at that. Without overtime compensation I averaged 60 hours a week for more than a year, and loved every minute of it. The Air Force project's design department at SOM was composed of aesthetically inclined types not necessarily grounded in the practicalities of building and so I retained my outsider status with them. But I was designing at last, and all those years learning about how to put buildings together helped enormously—George Fred Keck was right, after all.

During my first year at SOM, I worked under the direction of Walter Andrew Netsch, Jr., the respected, albeit self-important general partner in charge of design of the Air Force Academy project in Colorado Springs, Colorado. Ten years older than I was and another Chicago native, Walter's senior associates on the academy, Associate Partner Ralph Park Youngren and Participating Associate John Christian Hoops had both hired me as a junior designer, making me the "runt of the litter" of the rather considerable design team working on that very large project.

The Air Force Academy team was housed in separate quarters from the rest of the firm. We were located at 37 South Wabash Street in the heart of the "loop" across from the Louis Sullivan-designed Carson Pirie Scott department store. There was a singular camaraderie among us, and many of us respected the prodigiously accomplished job captains such as Carl Seaburg who could, seemingly without effort, detail their way through any construction challenge. He was but one of many whose constructional capabilities were amazing, principally to someone like me whose apprenticeship years were spent painfully attempting to accomplish what people like Carl could do in an apparently effortless offhand manner.

By the time I joined the design team in 1957, the academy project was winding down. The assignments that I received consisted for the most part of designing small-scale background support buildings, but I loved the neo-Miesian vocabulary that underpinned the architectural idiom associated with the work. In any case, the support personnel buildings were by necessity direct and to the point bereft of the highly ordered master plan and formally derived buildings with their aluminum cladding that marked the more significant academic and cadet quarters structures.

Working for Walter Netsch provided me with an inimitable post-graduate education in the politics of a drafting room—a kind of diplomacy I had never experienced before. Walter rarely criticized any of us explicitly. In my case, since Netsch had discovered that we were both die-hard Chicago White Sox fans, if he saw a design that I was developing that he didn't particularly like, he would introduce baseball into our conversation. The conversation would continue until I got the message that something I had done had gone awry, whereupon I would conversationally change course, embarking upon a self-criticism mode and proposing alternative ways to better the design other than the one I had shown to Netsch initially.

I suppose Walter felt that his method of criticism was a positive way to motivate the uninitiated, but instead this circuitous, manipulative style of criticism created a kind of resentment in me. Not unlike some other SOM general partners in the Chicago office, Walter rarely, if ever, gave his team members salary raises. Perhaps he thought that our connection with him was sufficient. Nonetheless, our relationship developed such that three years later when my son was born, Walter agreed to become his godfather.

Together with Ralph Youngren and John Hoops, I spent many nights at Walter's art-laden penthouse apartment on Lake Shore Drive working late on conceptual color patterning studies for the stained-glass windows of the Air Force Academy chapel. In sight of the magnificent Rocky Mountains, Walter's concept for the chapel was as an idiosyncratic "jewel in the crown" to the otherwise rational "neo-Miesian" design idiom reserved for the rest of the institution's campus buildings. I was pleased to help on this singular project precisely because my normal daytime assignments consisted of producing less expensive, watered-down versions of the same palette used for the officers and cadet quarters. Even so, these straightforward assignments became my opportunity to learn a language that was *de rigueur* for most other designers in the Chicago office.

By 1958 the Inland Steel building was ready for occupancy. It was commonly understood that Walter Netsch was credited with the basic schematic design of the building and that Bruce Graham was responsible for developing the concept, detailing it and bringing it to fruition. When the firm consolidated its offices and moved to quarters in the Inland Steel building, Bruce Graham the decisive, albeit dogmatic, 33-year-old former South American from La Cumbre, Columbia was made general partner in charge of design for the entire Chicago office. One of his first acts was to peremptorily dispense with the services of virtually all of the Air Force Academy design team. I was the sole survivor of the lot, in all probability because I was the lowest salaried. The blood bath was Bruce's way of exerting his authority over Walter, his former arch rival for the chief-of-design position.

Nonetheless, I loved working for Bruce. His conflict-driven, confrontational management style couldn't have been more dissimilar than Walter's. If Bruce didn't like something you were doing, he told you so in no uncertain terms. You never felt that you were being manipulated. If you disagreed with Bruce, and if you were bold enough, you could argue your case by engaging in a shouting match with him. Win or lose, at the conclusion of those disagreeable moments the air was always clear in the fifth-floor design studio situated at the north end of the Inland Steel Building.

It was like being back in the Navy. Like me, Bruce had initially served in Navy Air when he enlisted in 1942. In the service, when the person who outranks you issues an order, you in turn instruct the person who you outranked in a similar fashion. This kind of hierarchical chain-of-command system was something that I understood perfectly. To boot, if you demonstrated any sort of design capability, Bruce established allegiances by frequently giving his designers pay raises, bonuses and rapidly increasing responsibilities.

After the Air Force Academy project ended, one of my first assignments from Bruce was to assist Myron Goldsmith and Jim Ferris in the development of the Hartford Insurance high-rise office tower on the Chicago River in downtown Chicago.

After graduating from IIT and while on a Fulbright Fellowship in Italy studying with Pier Luigi Nervi, Myron was persuaded by his former IIT classmate and SOM general partner Bill Dunlap to return to the United States to join SOM, first in their San Francisco office and then in Chicago. Working with Myron I couldn't help but observe the painstakingly deliberate way that he studied large-scale elevations of the façade of the tower that, at least temporarily, slowed down my customary rush to judgment on issues relating to design.

Coming home from SOM at 9 pm, entering design competitions and producing paintings that I submitted for display at various art gallery exhibitions and neighborhood art fairs, I rarely got to sleep before 2 am. It was an invigorating time filled with many all-nighters at SOM, at least one of which was incredibly infuriating, and only in hindsight marginally amusing.

The incident occurred while our design team was preparing for a presentation of Netsch's signature Air Force Academy chapel to consultants Pietro Belluschi, Eero Saarinen and Walter's partner in charge of design in SOM's New York office, Gordon Bunshaft. A number of us had been assigned to labor over an enormous sepia reproduction of the original drawing that denoted the building's seminal transverse cross-section. Many changes to the design had been made by Netsch, and our task was to remove a large portion of the drawing with sepia remover, replacing the erasures with substantial changes he had made that we were to then execute in ink. Working throughout the night, we completed our work at sunrise, and just as we were preparing to pin up the replacement reproduction, one of our crew unintentionally spilled an entire bottle of sepia remover on the drawing, thus obliterating virtually the entirety of the original exclusive of the inked changes we had made. Needless to say, Netsch could barely conceal his rage at our clumsiness, but the consultants were far more sanguine about the entire episode, and our little band was thus able to put the incident behind us.

Like our team member who spilled the bottle of sepia remover, SOM had its share of employees whose personal habits were less than exemplary in the context of reasonable drafting room protocol; it's a malady not unknown to large-scale firms. For instance, one of our team members was a chain smoker who every so often set fire to drawings he worked on. When you came to work, you never knew what adventures would await you.

Other SOM designers were given to eccentric behavior that, at least in the studio of Walter Netsch, was deemed inexplicably acceptable. One of the designers on the Air Force Academy team appeared to be Walter's "art advisor" and as such was treated with kid gloves. He would arrive at the office late in the morning, take an extended lunch and leave in mid-afternoon, often going to gallery openings with Netsch. Considered by Walter to be a connoisseur of contemporary art, his guidance on matters of materials and color selection was sought on a regular basis. On one occasion, Netsch summoned his design team into his office to seek advice on the color of a wall axially terminating the interior entrance of a major support building. The "connoisseur" was given the task of selecting the color, thus ending the meeting. Four weeks later we were reassembled to hear that the absolute middle-gray color chip from Container Corporation's Color Harmony Manual had been chosen. Needless to say, employee overhead was not an important consideration in these proceedings.

*　　*　　*

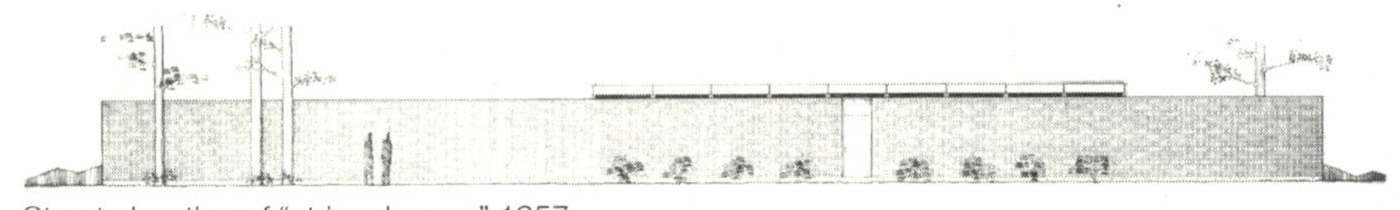

Street elevation of "atrium house," 1957

I began to moonlight, taking on projects to supplement my modest income at SOM. One of the projects, clearly attributable to Mies van der Rohe, was a one-story atrium house that, while it didn't get built, was published in its conceptual stage in the August, 1957 issue of *Arts and Architecture*—a first for me.

I didn't meet the magazine's distinguished publisher John Dymock Entenza until 1960, two years after he had come to Chicago as the first full-time director of the newly founded Graham Foundation for Advanced Studies in the Fine Arts. William E. Hartmann, the suave, well-mannered general partner in charge of SOM's Chicago office, had preceded Entenza as the part-time temporary director of the foundation until a permanent director could be found. Inexplicably, John Entenza was to become a kind of godfather to me, helping me immeasurably with my struggling architectural practice.

*   *   *

In 1961, after my return from Yale, I arranged an interview with John to discuss publishing work I had completed while a student. Our first meeting was unpromising. Typologically, a tall Spanish grandee, John Entenza's demeanor of noblesse oblige appeared condescending and unnecessarily cantankerous, and I told him so. Nevertheless, it marked the beginning of a warm 15-year friendship that softened my bull-in-a-china-shop persona, at least to the degree that it was possible at the time. John was unstintingly generous with his advice, and proactively recommended me for the type of commissions I would never have had access to otherwise.

John, a bachelor who lived in the same Mies van der Rohe designed "glass house" apartment complex as I did, often shared vodka martini cocktails with me on Saturday evenings, which was always accompanied by a slice of Hungarian salami rolled around cream cheese. John was an engaging raconteur whose stories entertained all who came into contact with him.

When John Entenza had a massive stroke while browsing in the second-floor architecture department of Kroch's and Brentano's Chicago bookstore in the late 1970s, I hastily joined a trio of his influential claque of WASP dowagers who had gathered by his side as he rested on his gurney in the emergency room at Northwestern Hospital. I had been summoned by a call from John's friend Henry Tabor, the head of the architecture section at Kroch's and Brentano's. Among other disabilities brought on by his massive stroke was a severe case of aphasia.

John had lost the capacity to speak, but somehow I could understand the few words that he was attempting to exchange with me much in the way that people who are close understand each other in similarly calamitous circumstances. When John tried to convey his thoughts to me in the hospital emergency room by appearing to ask through eye contact "what was happening architecturally" at that very moment in Chicago, I somehow managed to understand what he was trying to

The author (left) and John Entenza in La Jolla, California, 1977 say. When I responded that the New York architect Peter Eisenman was in town just at that moment to give a lecture, John struggled to express himself as coherently as he could, ultimately uttering a single, appallingly crude anti-Semitic word that stunned his WASP dowagers, but was really only a representation of his inability to control himself. Strokes sometimes have that effect on their victims. In any case, I never took offense at his utterance.

Before I married my third wife Margaret McCurry in 1979, we settled John in at Casa de Mañana in La Jolla, California. Only in Southern California would the natives be either so world weary or so insensitive as to name a retirement facility "House of Tomorrow." Slowly, as he regained some limited speech, John always insisted on referring to the La Jolla facility pejoratively as his "Presbyterian lock-up."

After its Chicago debut and the subsequent exhibition at Minneapolis' Walker Art Center, our *Chicago Tribune Competition* redux exposition opened at the Museum of Contemporary Art in La Jolla, whose director at the time, Sebastian (Lefty) Adler, had a record of supporting architectural extravaganzas. Lefty and I dedicated the opening ceremonial event to John. When I picked him up at his retirement facility to take him to the opening celebration, John was schmoozing as best he could with an old lady, who upon seeing me, observed to John, "How pleased you must be to have such a doting son!" The stroke had robbed John of the nuances of speech that might have allowed him a suitably scathing riposte, but he was not beyond acting out his frustration by slapping his hand to his forehead, and exclaiming: "Oh, God!" Soon thereafter, he had two subsequent strokes, the latter of which was massive and he succumbed, robbing our community of one of the most civilized men in our field.

*   *   *

Before he was felled by his first stroke and as he was approaching his retirement as director of the Graham Foundation, John Entenza and I were among a small group who observed an incident that transpired at the foundation following an interminable (as usual) lecture by John's old friend Buckminster Fuller. At the reception, John introduced Bucky Fuller to Charles Murphy, Jr., whose firm had just completed the much admired Chicago Civic Center, and whose father Charles Murphy, Sr. had started the Graham Foundation.

Fascinated by the long-span high-rise structure that had been designed by Jacques Brownson of the Murphy organization, Fuller asked Murphy not what the building cost per square foot, but what it weighed per square foot. Murphy's puzzled expression told the story; he had never perceived a building in those terms, whereas Fuller, not trying to put down Murphy, merely wanted to ascertain the building's value so that he could measure it against other structures built in a like fashion. The question, asked several decades before people would find such observations meaningful, was a marvel of inquisitiveness and a hint of what could conceivably transpire in the not-too-distant architectural future.

It also made me think that as buildings became lighter and possibly less static, and in any case more permeable, someone would surely write a "History of Architecture by Weight." Buckminster Fuller's comment seemed to imply that the future of architecture might well be about plastics and adhesives and that particular future was upon us at that very moment. Already Fuller's built structures utilizing triangulation, combined with "least surface" experiments by the brilliant German engineer Frei Otto (who was to be my studio critic later at Yale) as evidenced in his tensile-tent structures, were having a significant impact on the way that many of us began to think about architectural structure.

I was privileged that John Entenza took such a keen interest in my emerging career and offered such strong support over many years. Esther McCoy, the well-known California architecture critic, was only one of many of his friends and colleagues who he persuaded to publish my work both in the United States and abroad. If there was one person in my life, proving the case that "no man is an island," it was John Dymock Entenza. In his will, he left me his 2,000-book architectural library, now dispersed throughout my own 5,000-book collection. He was an intellectual giant whose erudite persona in these intervening years has been sorely missed in a city not normally known for producing people who provably had perceivable IQs.

*   *   *

In the context of professional egocentrism, I attended a symposium in 1958 at Chicago's lakefront convention center (the first McCormick Place, designed by Edward Durrell Stone, which was later destroyed by fire). The conference was to feature Mies van der Rohe's residential developer-patron, the same Herbert Greenwald I had met nine years earlier, and I.M. Pei's developer-client William Zeckendorf, both men inexplicably sharing the stage with Frank Lloyd Wright. Mr. Wright was the last of the three to speak and as he waited impatiently for his turn, he sat on the platform snoring while feigning sleep. Alternatively discourteous and haughty, he interrupted the other two speakers at will. When it was finally his turn at the podium, Mr. Wright was abrupt and to the point, stating that, "given present company [he] had no business being there." With that, he departed the auditorium, leaving his audience stunned and silent at first, and then indignant.

But that was not my first exposure to Frank Lloyd Wright. Three years earlier, in the summer of 1955, nine months after I returned from military service, Don Anderson, a local architecture critic, and I drove up to the school/ studio that Wright ran at Taliesin in Spring Green, Wisconsin. Neither of us had been there before, so on a beautiful Saturday afternoon in June, we drove northwest from Chicago in my little second-hand Morris Minor that I had purchased from my commanding officer in Norfolk, Virginia during my last year in the Navy. Our conversation in the car ranged around the cult status that the descendants of the greats placed upon those who they idolize.

We arrived at Taliesin just prior to 6 pm and, purely by chance, parked the car in the vicinity of the entrance to the dining pavilion. As we walked the few steps toward the building, the clock struck the hour and we were startled to hear a J. Arthur Rank-like gong go off. Halting in our tracks, we couldn't help observing a Cadillac convertible with its top down pulling up to the curbside. Mr. Wright, clad in a dark cape, and his wife Olgivanna, resplendent in a flowing white gown, regally descended from the vehicle. Just like it is customary in the Navy when flag-rank officers are piped aboard ship, "side boys" quickly appeared wearing plain white T-shirts, white duck trousers and sneakers. The young men flanked the walk, while a youthful maiden clad in a white gown (I am at a loss to describe her in any other manner) strewed white rose petals on the sidewalk before the two principals in that particular passion play as they deliberately made their way in to dinner.

Dumbstruck by the cult-like processional that we had just witnessed, Don and I nodded knowingly to each other and resolutely made our way back to my pocket-sized vehicle. As we hurriedly motored back to Chicago, we noted that we were quite possibly the only architecture buffs who had ever spent less than fifteen minutes at Frank Lloyd Wright's idyllic retreat at Taliesin.

*   *   *

Meanwhile, the upside of the Chicago office of SOM was the quality of the work itself and the collegially open spirit that straightforward teamwork based upon mutual respect produced. In that vein, Harvard-trained Robertson Ward, Jr., SOM's brilliant resident materials research and development innovator, Associate Partner Edward Petrazio and I jointly developed a cruciform stainless steel column whose antecedents could be found in the columns of Mies van der Rohe's Barcelona Pavilion. Even though we knew that the intrinsic shape of the column had extreme fiber bending limitations, we nonetheless had stainless steel extrusion prototypes fabricated by the American Bridge Division of United States Steel. While the column was never manufactured, the three of us developed the extrusion without any thought to individual credit.

The downside of SOM was the political machinations engaged in by some of its most senior members. One of the original firm founders, Nat Owings, reveled in the power struggle between Walter Netsch and Bruce Graham, which in due course became ruinous to their health even as it became unnecessarily wearying on the entire office. These office politics exacerbated my feelings of being unfulfilled brought about in part by my lack of academic credentials. I also felt that the hotbed of ideas produced at a lively architecture school might reinforce my desire to establish a raison d'être for what I designed. In the midst of this and when I least expected it, a happy accident transpired at the annual SOM Christmas Eve office party of 1958 that would redirect my life forever.

Sitting in Bob Ward's office during the party, I indifferently leafed through a recently published book from MIT entitled *The Curtain Wall* that was lying on his desk. It was authored by a claque of graduate students at MIT's architecture school. I observed that while some of the authors' undergraduate degrees were, as expected, noted in the acknowledgements, others were not. I later came to find out that the unlisted undergraduate degrees were nothing more than an inadvertent printing error.

Never mind, did that mean that I could attend graduate school for one year and get a master's degree without benefit of an undergraduate degree? This pursuit could conceivably occur in an environment that was intellectually and aesthetically driven.

Innocence ruled the day. In typical Tigerman fashion, without further ado or "due diligence" and tipsy from more than the alcohol I had consumed, I excitedly telephoned my wife Judy from the SOM party to ask what she thought about my going back to school that very next autumn. The following day I sent letters off to Harvard, IIT, MIT and Yale outlining my ambitions for graduate school and requesting applications to each of their graduate architecture programs.

Some years later I discovered that in the late 1950s the application form to Columbia University's graduate program in architecture required answers to three questions that were, in hindsight, simplistic yet extraordinary. The first asked the applicant if he or she would design a house for Fulgencio Batista, then Cuba's dictator; the second asked if the applicant would design urban housing at a density greater than what was commonly acceptable at the time; and the third asked if the applicant would design a concentration camp. Simplistic perhaps, but ethically derived, certainly.

In any case, in response to my queries MIT was the first to react, their letter virtually arriving by return mail. Their note stated that they had reviewed my decade-old student transcript and as a result, they would be happy to reaccept me back into their undergraduate program for the prescribed four more years culminating in a BArch degree. I tore that one up. Harvard's reply suggested that while I sounded good, they had their rules. I tore that one up as well. I have yet to hear from IIT.

Anxiously, I went through my mail every day, waiting in vain for any response whatsoever from New Haven. Two months passed before I received a handwritten note penned by 39-year-old Paul Marvin Rudolph, the celebrated wunderkind architect and controversial chairman of the Yale School of Architecture who wrote, "I know I will live to regret this, but please find enclosed an application." I responded by return mail, sending a care package to New Haven of examples of my drawings and, without further ado, anxiously boarded the first flight out of Chicago heading east for an interview at Yale.

As a prelude to my meeting with Mr. Rudolph, I first had an appointment with Yale's distinguished but crusty New England Yankee, the eminent architectural historian Carroll Louis Vanderslice Meeks. His well-known book *The Railroad Station: An Architectural History*, first published in 1956, thematically presented 19th- and 20th-century architectural history through the lens of the train station. He seemed amenable enough when he heard about my wish to return to school, but I had yet to meet with the person who would make the decision, The Chairman.

My interview with Paul Rudolph in the intimidating, neo-Gothic Weir Hall facing the courtyard delimited on one side by Louis I. Kahn's Yale Art Gallery and on the other by Jonathan Edwards College did not start out on an especially promising note when he learned of my intention to acquire a master's degree in just one year with no more than a single uncredited year of college behind me resulting from my flunking out of MIT a decade earlier. Uncharacteristically for the normally short-fused Paul Rudolph, he patiently informed me that the master's degree in architecture at Yale had a prerequisite of the bachelor's degree in architecture which I needed before I could matriculate into the master's program.

I obstinately reiterated that, given that I was already a registered architect, I wanted to obtain the master's degree in but one year without qualification. Rapidly losing patience, Mr. Rudolph further observed that Yale's primary architecture degree (BArch) required a liberal arts undergraduate diploma (BArts, or BSci) to get into graduate school in New Haven in the first place. Digging my heels in, or not paying attention—I am not for the most part noted for my powers of concentration when my obsessions surface—our now embattled negotiations began to deteriorate dramatically.

Finally, perhaps out of exasperation combined with a desire to bring those ostensibly unproductive discussions to a close, Mr. Rudolph proposed a non-negotiable compromise. He would allow me to matriculate into the bachelor's thesis year whereby I would potentially acquire a B.Arch degree in one year. Assuming that I excelled in the bachelor's thesis class, he would further allow me to join the subsequent year's master's program thereby acquiring the M.Arch degree the second year.

Elated, I exited the meeting before Paul Marvin Rudolph could change his mind, and by a strange twist of fate returned to New York on the same train with him. Putting aside our earlier tense negotiations we shared a pleasant trip together; one of several that we were to jointly undertake a half-decade into the future.

*   *   *

Returning to Chicago and my post at SOM, within months I discovered that thanks to the expertise of Southside baseball hitters and fielders like Luis Aparicio, Larry Doby, Nellie Fox and Sherm Lollar and low ERA pitchers like Billy Pierce and Early Wynn, my beloved Chicago White Sox were American League pennant contenders. I laid down a series of bets against a number of unsuspecting aesthetes in SOM's design department that netted me more than one thousand dollars (half the Yale tuition at the time) when my beloved "boys of summer" won the 1959 American League pennant. I gathered up my ill-gotten gains, packed our secondhand stick-shift Ford Falcon station wagon with what little we owned and headed east.

*   *   *

Two years in New Haven would seem like a lifetime. First off, my wife and I lived in three locations in just two years that required unnecessary time spent in moving from place to place. Yale's architecture school under Paul Rudolph's direction was perceived by students as existing somewhere between the Marine Corps Base Camp Lejeune's drill-training boot camp and John Houseman's excruciatingly matter-of-fact film, *The Paper Chase*, depicting the rigors of Harvard Law School.

Although I was a registered architect, my negligible academic credentials did not include any of Yale's required architecture support courses such as structures, environmental systems, professional practice, et al. However, to my surprise, Paul Rudolph was flexible with respect to my proposed course selections. With his concurrence I looked forward to taking advantage of a distinguished university's breadth of course offerings that could endow me with some level of enculturation that I missed in the lost decade that preceded my coming to Yale. But first I had to gain acceptance into those courses by the appropriate faculty.

Before I was privileged to burn up countless hours on Josef Albers' celebrated color course (he was by that time a professor emeritus and it would be the final year that Professor Albers would teach that

95

particularly rigorous class), I had to get his consent to enroll in his studio. He was disinclined to allow me entry into his class, stating that he was apprehensive about architects, since those specifically under the sway of Mies van der Rohe had a tendency to design furniture by squaring off one's "derriere." After Moholy-Nagy, apparently Professor Albers also had problems with Mies van der Rohe's educational policies at the Bauhaus in Berlin. Even so, he overcame what misgivings he had about me, grudgingly allowing me to matriculate into his sought-after course.

I took a cornucopia of classes in addition to Josef Albers' color course. George Heard Hamilton, the curator of *The Société Anonyme*, delivered brilliant modern art history lectures. Paul Weiss challenged undergraduates with his metaphysics seminar and the artist Bob Engman taught a graduate sculpture studio. Those courses combined with Paul Rudolph's architectural design studio for the bachelor's thesis class were enough to remind you just how precious sleep was. In that pressure cooker, if you weren't continuously competitive, there was no way that you would live to tell the tale.

I learned that lesson immediately and at my own expense. Paul Rudolph opened the fall semester of the bachelor's thesis class in September, 1959 with a one-week-long sketch problem. The assignment was to design a mid-block infill luxury urban townhouse situated on Manhattan's East side on Beekman Place. Fresh from Chicago and Mies-influenced SOM, I effortlessly cranked out what I thought was a well-designed minimalist plan and presented it in a crisply collaged Miesian arrangement.

At the subsequent jury that included the current visiting architecture faculty imports, the Brutalist architect Jim Stirling from the UK and the watered-down neo-Miesian Craig Ellwood from LA, everyone appeared to respond positively—if not enthusiastically—to my design solution. Regrettably in my case, when the grades were posted the next day, I was shocked to discover that I had failed!

Crushed, I walked home at a snail's pace that warm autumn night in tears. Humiliated, I thought about leaving Yale, only to slowly come to the realization that Paul Rudolph was not having some smart-ass "registered architect" breeze into town, trumpeting his drafting skills. "Toughen up" was the cathartic message I sensed that he had conveyed to me, and only then did I understand that I had to buckle down and do the very best that I was capable of accomplishing if I was to thrive, not just survive at Yale. That decision meant putting aside whatever presentation skills I had learned in Chicago from my employment by second and third generation Miesian descendants, expending energy more on ideas than "eyewash."

That's the way Paul Rudolph ran the school. For example, after several weeks of getting grades of

90 or better in the bachelor's thesis design studio working on a project directed by Jim Stirling, I was distressed to see that my grade for the most recent week had slipped to 85. There was a terse note in my student mailbox from the soft-spoken African American, southern-born registrar Hazel Nelson, requesting my presence for an interview with The Chairman.

At the appointed hour I trudged across Weir Court from Lou Kahn's art gallery expecting the worst. I was not disappointed as I tremulously appeared before Mr. Rudolph in the cold, Gothic, dressed stone interiors of Weir Hall. In turn, Mr. Rudolph introduced me to a mysterious Ivy Leaguer replete with crew cut, blue blazer, rep tie, chinos and white bucks who was in attendance. I was informed that "Mr. Smith" (or whatever his name was) was associated with the Yale placement bureau. Referring to my last week's "substandard grade," Mr. Rudolph then said that I seemed to have lost interest in architecture. However, he told me that the architecture department would gladly undertake the responsibility of underwriting the cost of a battery of tests to ascertain what field I would be more suitably qualified for other than architecture.

I distinctly remember doing some sober groveling that included meekly pleading my case and swearing somberly that if I were but given the opportunity to mend the error of my ways I would do better in future. An embarrassing experience, but that seemed to be the point of the exercise.

*   *   *

When exasperated to the point of losing control, Paul Rudolph stuttered. A number of his students, notably Charles Gwathmey and yours truly brazenly emulated his malady while presenting our projects on juries. We hoped that our stuttering would infer uncertainty so that we would appear to be modestly disposed towards our work without actually mocking The Chairman. For me, it simply meant reinstating my much earlier stuttering habit. While I have the feeling that Rudolph was to some extent amused by such antics, Gwathmey and I were playing a high-risk game and we knew it.

The three students at Yale with whom I was closest were Charles Gwathmey (two classes behind me), Jacque Robertson (one class behind) and Bob Stern (three classes behind)—this was decades before he became Robert A.M. Stern. The three of them represented a fair cross-section of students at the architecture school at the end of the 1950s.

Charley Gwathmey's architectural production, both as a student and later as a practitioner, much like his well-honed muscular physique, had always been grounded in matter-of-factness, even as he was influenced by the aesthetic antecedents of Le Corbusier. Jacque Robertson has always been the penultimate Yale-trained, Oxford-pedigreed, Virginia-gentleman architect in his approach to architectural education and through the vehicle of his classically inclined work both as a planner and as an architect. He expressed his sophistication and connoisseurship in his rich baritone voice with his ability to converse in complete sentences. Bob Stern was the only student who had the temerity combined with an embarrassingly massive dose of chutzpah to address The Chairman as "Paul;" the rest of us would never have been caught dead referring to The Chairman by his first name. Gwathmey (until his untimely death in 2009), Robertson, Stern and I remain friends to this day.

As students we were close, but not too close, as one's competitive streak forbade excessive intimacy. For example, immediately following the fall semester's final jury representing an office-building project in New Haven on which the participants in our thesis class labored, and immediately prior to the 1959 Christmas break, Rudolph handed out his program for the sketch problem that was due on the very day that we were scheduled to return from our holiday. Vacation plans were made and plane reservations were in hand when one member of our bachelor's thesis class implied that he might stay a few hours to review the design problem before leaving for vacation. The result was that competitiveness combined with insecurity caused everyone to curtail their holidays. While the rest of Yale traveled to someplace like the Bahamas, or wherever preppies vacation, our studio on the fourth floor of the Yale Art Gallery was fully occupied, Luxo lamps aglow over our drafting boards all December long.

*　　　*　　　*

We were all in a crucible, slowly simmering, while trying to avoid being scorched, and yet there were some surprisingly Shakespearian moments of comic relief. One such memorable moment occurred when those of us with our drafting boards adjacent to the glass wall that faced Weir Court observed that one of the students in the school was "having his way" with a female art student. The pair was coupling in the valley of a dormer on the gable roof of Jonathan Edwards College, which faced the architecture school just across the courtyard. What on earth prompted them to engage in such intimacies in such a public place remains unknown. Removing the inside screws that held one large pane of glass in place, Charley Gwathmey single-handedly lifted the glazed section out of its position and a goodly number of us cheered our friend on with three "hip-hip-hurrahs" before the two of them, red-faced and mortified at being discovered, hurriedly retreated back through the dormer window to the relative privacy of Jonathan Edwards.

High-style moments in New Haven appeared when least expected. The theme of the annual Beaux Arts Ball that year was "Black and Blue." My wife Judy and I thought we had clever costumes when we dressed as bandage-clad casualties but we were small potatoes next to the winners of the best costume award that year. One of the members of the master's class had dyed his ample body and hair royal blue arriving at the party au natural with his equally unclad statuesque African American date.

Another such moment came when we all observed our classmate Phyllis Lambert, the Seagram heiress, come late as usual to a jury dressed to the nines like the occasion was a smart cocktail party. Closely behind her strolled her Bahamian man-servant balancing her architectural model on his extended fingers like it was a tray of canapés. Equally entertaining was watching the same Phyllis Lambert sauntering down Chapel Street on her way home to her refurbished town house on Crown Street with her chauffer-driven Rolls Royce Silver Wraith following slowly at a discrete distance behind her.

Less amusing was her inopportune feminist habit of refusing to walk down one flight of stairs to the women's restroom. Instead, when nature called, she would frequent the men's facilities that were adjacent to the fourth-floor drafting studio. She would simply barge in without so much as "excuse me" and deliberately make her way to the water-closet stalls at the rear of the restroom. This bothersome behavior caused many the male to cringe causing stained chinos for those of us who had the misfortune to be utilizing the urinals as she nonchalantly sauntered by.

All our juries seemed like public events. More often than not, they were populated with passing New Haven civilians as well as homeless folks from the Yale Hope Mission who happily joined in the wine and cheese refreshments not paying particular attention to the proceedings.

At one unforgettable jury the son of one of the founders of a major architectural firm, presented a transparently Le Corbusier-informed project that was extremely reminiscent of Corbu's monastery chapel at Ronchamps. Paul Rudolph challenged the student's choice of precedent by noting that his father's firm's projects had up until that very moment all been influenced by Mies van der Rohe, but as Rudolph noted, "that was okay" because he knew where the student had come from and, after his stopover at Yale, where he would no doubt return.

When the young man suggested that he had accomplished his goal "of learning everything there was to be learned about Mies," and that he wished to move on to study the work of Le Corbusier, red-faced Rudolph retorted that the luckless student had learned little about Mies, but Rudolph would give him the

The author with the model of his apartment tower, 1960

The author presenting his bachelor's thesis project, 1960

opportunity to do just that by removing him for a time from the school such that he would be forced to repeat the studio. The assembled body applauded, but at the same time understood that any of us could at any time be subject to similar brutality.

The pressure of the place was such that while it sounds amusing a half-century after the fact, the sobering reality was anything but. Even so, the nature of survival seems to qualify one to make some mockery of momentary glimpses into a Yale architecture student's life.

My own bachelor's thesis jury became a memorable experience from several points of view. My presentation included 16 20-inch by 30-inch boards that I had drawn immaculately in ink together with a 6-foot-high three-dimensional representation of my apartment tower. My proposal was both drawn and modeled at a scale of one-quarter-inch equals 1 foot and was topped off with an original Bob Engman topologically informed bronze sculpture. The model was perched perilously atop four drafting stools in the entrance lobby of Weir Hall hovering over all who witnessed the event.

Simplistically, the apartment tower that I proposed was formally organized in a "pinwheel" configuration, composed of prefabricated concrete cubic forms rotated in a rectilinear manner about a structurally defined central utility/vertical circulation core from which they were anchored.

The jury, typical of Yale juries then and now, was populated by the architectural luminaries of the day, one of whom was the same cantankerous Serge Chermayeff with whom I had been at variance with during my Institute of Design days a decade earlier in Chicago. The jury was virtually undivided in its praise for my project, when Serge Chermayeff unexpectedly arose, suggesting acerbically that: "between us, if you had it to do all over again, you probably wouldn't have designed it the way you did." I was never given the chance to give explanation for what I had done. Rudolph at once jumped to his feet and contemptuously responded: "Of course he would, you ass: next project!"

*   *   *

Many of my Yale colleagues felt that the reason Paul Rudolph brought in so many "architect luminaries," many of whom held views that conflicted wih his own, was to tax their architectural belief systems against his own strongly held views by publicly testing them. Rudolph's attitudes about space and mass reflected his fascination with Wright's and Corbu's use of perspective and the drawings those architects made that reflected their perceptions about "deep space." Those of his guests who didn't share those beliefs were often the recipients of cutting commentary. He wasn't bashful about sharing his belief systems within earshot of anyone and arguing vociferously on behalf of his position.

After sarcastically and explicitly challenging his guest jurors, Rudolph reformatted himself by becoming the picture-perfect host at the parties that he threw after juries and lectures. These fetes were held at his personally remodeled duplex apartment on High Street around the corner from the school. His generous double-height living accommodations bore the stamp of his idiosyncratic interpretation of modernist domestic interiors. The soaring living room opened onto an idyllic rear garden that was a background to the theatrical quality of the interior. Mr. Rudolph's professional architectural office was on the top floor of the very same building.

For my own bachelor's thesis jury at the conclusion of the 1959–60 academic year Gordon Bunshaft, Jean Paul Carlhian, Harry Cobb, Ulrich Franzen, Philip Johnson and James Stirling were among the well-known architects of the day whose presence was requested at Rudolph's obligatory spring celebration.

My wife Judy, who had never previously been to Rudolph's stunning two-storey apartment and who was understandably intrigued by its unique modernist design, wandered throughout the upper-entry story, snooping about. She opened one door after another until she opened the unlocked door to the bathroom, to find the prominent architectural historian, Vincent Scully, perched upon the throne. Excusing herself, she hurriedly descended the cantilevered marble and white-painted steel stairs to the double-height living room below, where she undiplomatically announced her awkward discovery to the assembled body as they gathered around the grand piano. Hearing her tale, the group burst into laughter at Mr. Rudolph's observation that, "they had all waited for a long time to catch Vince with his pants down."

Professor Scully never returned to the party. The following year, whenever Vince Scully and Judy's paths were to coincide on campus—she was the secretary to the dean of the law school at the time—he would cross the street to circumvent making contact with her.

In any case, Rudolph had a distant relationship with Professor Scully during the time they shared at Yale, as he did with so many others. His determination to shine as a bright star in the architectural firmament precluded deep friendships within the field. As for Scully, the historian was far more a supporter of Louis Kahn's architecture than he ever was of Paul Rudolph's. Immediately upon Rudolph's resignation in 1964, which became effective in 1965, Scully changed horses. The ensuing appointment of Rudolph's exact opposite as his successor, the postmodernist-minded Charles Moore, also brought Robert Venturi and his wife Denise Scott Brown to the campus. Scully then helped to elevate Venturi's reputation within the architectural community, thus making the strained relations between Rudolph and Scully somewhat more enduring.

*   *   *

My proud parents, who finally saw their dreams for me fulfilled, attended my Yale graduation in 1960. With eyes moistened with the poignancy of the occasion upon me, I handed my mother the degree that had taken me so many years to obtain. Her immediate response was, "Now you can make money!" True to form, her never-ending longing for that which had always eluded her surfaced once again.

It seems that our contradictory priorities remained unchanged. Nor did I ever have or take the opportunity to attempt to amend her life-long ambitions that she repeatedly tried to reassign to me. That I did so little to break down the barriers that separated us on this sensitive subject troubled me no end. Eighteen months later she was dead, and I still harbor the remorse associated with how little I did to communicate with her the nature of my personal architectural aspirations.

My mother had spent her adult life complaining of an assortment of ailments that never seemed real to me, yet she died of the aggregation of all of them. Although never close to my mother in conventional terms, I was nevertheless crushed at her death, perhaps as much by the state of our unresolved relationship as by her unforeseen departure from my life and this earth that we so tentatively shared with each other.

A decade later my father, who had subsequently married Rose, a woman who not only looked remarkably like my grandmother but shared her first name as well, died after a massive stroke. His lifelong sweetness masked an uncomplicated, yet unfulfilled life. As a middle child he was bracketed by stronger brothers, but in my view his gentle qualities far transcended any significant characteristics one could attribute to his siblings. As much as I loved my father, it was my mother who was unquestionably my alter ego. I inherited her obstinacy and many of her preconceptions; we were alike in too many aspects of our individual personae for me to ignore that in turn led to any number of conflicts between us.

*   *   *

After the Yale graduation ceremonies, Mr. Rudolph invited my parents, my wife Judy and me back to his marvelously dramatic apartment. My mother who was burdened by bifocal glasses cautiously picked her way down those notoriously cantilevered steel-framed stairs with white-marble treads but without a handrail. At the bottom, ebulliently ensconced in the living room

Joe, Sam and Milton Tigerman, 1965

"pit" and with the intention of flattering Mr. Rudolph, mother told him what a good husband he would make. Observing his stuttering hesitation to respond to that comment, she elaborated further, "but, you do like girls, don't you Mr. Rudolph?" whereupon immediately I did everything in my power to conclude that conversation and to get everybody concerned out of Rudolph's apartment as quickly as possible before any further "Jewish mother" type bons mots came out of her unpredictable, but astoundingly discerning, mouth.

*   *   *

After a summer back at SOM, in my second and final year at Yale, in addition to the few extracurricular hours that I spent daily on the fifth floor of the Payne Whitney Gymnasium assisting the fencing master Andre Grasson in coaching Yale's varsity saber team, Paul Rudolph hired me to work in his office after the architecture school closed for the night.

In those years, the architecture studio terminated its daily activities promptly at 2 am when the Yale radio station blared out the Yale anthem "Bright College Years." Every night in an unanticipated explosion of collegiality, we all rose from our drafting-board stools to belt out Yale's alma mater. Many of our group then reconvened at My Brother's Place, the local architecture-school hangout on Chapel Street across from the art school at Street Hall. I instead went to work at Rudolph's atelier around the corner on High Street until 5 am.

Working in Paul Rudolph's office was an eye-opener. For those of us who thought that we had a strong commitment to architecture, what we engaged in was child's play next to Rudolph's personal work ethic. He toiled tirelessly night and day, intermittently striding the one short block from his studio to the architecture school whenever the spirit moved him. One never knew when he would show up on the fourth floor of the Kahn building, but God forbid if one's drafting board was unmanned when he arrived.

For someone like me, who had his own fanatical approach to architecture, observing Rudolph laboring in his garret-like studio at his drawing board underneath white-painted wood warren truss rafters was truly inspirational. To produce a perspective rendering for example, he would begin sketching out the drawing starting at the upper left-hand-corner. Sometimes, those drawings were prepared without the benefit of any preliminary penciling. Like the way in which New York self-endowed cognoscenti daily address the *New York Times* crossword puzzle, Rudolph treated the vellum sheet on which he worked as if it were a matrix attacking it diagonally and completing the drawing expeditiously.

Working in Rudolph's studio was a race against time for the rest of us as well. I thrived on it, but there was a price to pay. I was frequently in a state of sleep deprivation and I looked it. No matter, I was at my drawing board in the architecture-school studio every morning when it re-opened at 9 am, as were all of us.

The pressures connected with keeping up with the demands of the school were only part of the problem. Money was very tight, and even though I had a full scholarship, there wasn't a lot of fungible cash on hand for such essentials as, say, food. I became expert at irregularly showing up in dissimilar apparel to maintain anonymity at cafeterias around town thereby eating free meals—ketchup and hot

The author in his standard sleep-deprived state, 1960

water passing as tomato soup with mustard and cracker sandwiches washed down with water.

The forces exerted on architecture students were such that each of us needed some outlet lest we explode. In my case, it was more literal than not. During the last winter of my final year at Yale, three of us in the master's class decided to play a prank by recklessly exploding a small device, a bomb, in Weir Court. The device was sufficiently powerful that when we were brought before Dean Gibson Danes, he came right to the point when he informed us that he didn't want to see us again until graduation—or we wouldn't. Frank Gencorelli, David Samuel Scheele and I sheepishly retired to our drafting boards and tried to maintain inscrutability for the rest of our final springtime in New Haven.

That major aberration aside, I developed a shared spirit that remains a distinct part of my personality to this day. It surfaced for the first time at the beginning of the 1961 spring semester when I collectivized other architecture students to discuss, and/or to present their ideas and their projects. Carrying that interactive spirit further, I contacted contemporaries that I knew from Harvard and Penn to be present at a weekend conference I organized at Yale during which we debated the issues of the day as we perceived them. The bad news is that not much that was informative or even useful surfaced, the good news is that the participants, and by extension their respective institutions, developed an intercollegiate camaraderie unexploited before that time. One result was that other similar events came about as students began to feel collaboratively inclined.

*   *   *

Paul Rudolph was easily the most demanding teacher I ever had. Even so, his expectations were no more rigorous than those that he placed upon himself. He had incontrovertible standards which needed to be met concerning architectural production, yet he was surprisingly sympathetic to the problems that student architects faced such as exhaustion, lack of funds, deteriorating marital relationships, et al. My own matrimonial state of affairs had begun to unravel since my arrival in New Haven and was further exacerbated by the grueling regimen and considerable amount of time away from home required to stay the course.

Rudolph had little time for fawning students, although there were more than a few of those at the school who thought that by emulating his work they could somehow survive the trial by fire instituted by his scrupulous management style; no such luck. If you thought that pandering to Paul Rudolph's stylistic predilections by making them your own would confer upon you credibility in his eyes, you were in for a rude awakening. Whether or not your capabilities were effectively expressed in your architectural production, Rudolph would eschew your stylistic predilections as he criticized your work so as to help make it the best that it could be on your own terms, not his. He wasn't shy about informing you of your strengths and/or weaknesses in the context of design. That straightforward approach to criticism was, at least in my case, eminently useful in organizing my thoughts about architecture. Brilliant yet brutal, his critiques unearthed fires that I never realized were burning.

Paul Rudolph also exhibited little patience when his student's significant others attempted to lay a guilt trip on him about how loyal they were to their partners. In early December, 1960 my wife Judy was in the design studio helping me with my presentation drawings by dotting in grass in ink on a plate of mine. As she was within a week of giving birth to our son, her arms were fully extended over her distended form as she was draped over the drawing board. When Rudolph passed by, she pointedly noted her condition only to receive his immediately delivered unsympathetic response, "but that is your job!"

On a cold winter night a week later our son was born and because of a less-than-attentive Yale medical school obstetric intern, Judy went into a coma, almost dying at Grace New Haven Hospital.

*   *   *

Since Rudolph tended to rebel against any and all regulations imposed upon him by anyone or anything, he was the bane of Yale's provost. Students who got straight As in all their courses other than design, because Rudolph believed they were incapable of expressing sufficient design capabilities in the studio, were lucky to get their degree on time, if at all. On the other hand, if students flunked most other courses but Rudolph felt that they were dazzling designers, the chances were that such persons would not only graduate on time but would be awarded one of the several traveling fellowships that were available to graduating students who had excelled in design.

In the 21st century, given the inflated costs of education combined with the very real threat of litigation initiated by disgruntled parents, it is the rare architecture student who flunks out of school. That was not the case in Paul Rudolph's seven-year tenure as Yale's architecture school chairman. For example, only 50 percent of my own bachelor's-thesis class graduated on time. Some did remedial work over the summer, others took an extra semester or year to complete their work and still others never did receive their degree.

Even so, the brutal rigor of the place was every now and then ameliorated by some ironic incident. In the spring of 1960, when a relic from an earlier class presented his bachelor's thesis plan alongside our own thesis class work, Mr. Rudolph was disgusted with the student's effort, cruelly asserting that he considered the project well below par for the course. He acerbically assured the student publicly that, "as long as I am chairman, you will never get the BArch degree!" Interested in that ultimatum, I followed that student's progress in subsequent years, as the luckless creature dutifully presented thesis project after thesis project, year in and year out.

Nineteen sixty-one, 1962, 1963, 1964 and 1965 and still the student flunked his thesis jury. Finally, Paul Rudolph resigned the chairmanship of the architecture department and in 1966 the student presented yet again. Since there was no Paul Rudolph to thwart his ambitions, the student finally acquired a passing grade for his efforts and of course his long awaited degree. Every so often obstinacy wins the day.

*   *   *

Little "Rudolphians" had a far worse time of it at Yale in the 1960s than "Miesian sycophants" did at the Illinois Institute of Technology where they were celebrated by a moribund caretaker administration. By 1959, IIT's president John Rettaliata had fired Mies van der Rohe who was subsequently replaced by acolytes who tediously regurgitated what was a proven stylistic preference for several successive decades.

Surviving Rudolph's cathartic cauldron required struggling to unearth resonant solutions to design problems with which individual preconceptions had little to do by digging deep into the reservoir of one's imagination. Curiously, not much has been written about that unique aspect of Rudolph's educational legacy, but his ability to critique others on their own terms while still requiring considerable amounts of work to flesh out their design was, for many of us, the most significant aspect of his tenure as a studio critic at Yale.

For practicing architects who also choose to teach, it is certainly easier to profess what is intrinsic to their own work rather than to approach the student on an open and level playing field. In my own case, Rudolph's demands as my studio critic in combination with his flexibility about what I was trying to express through my work resonated with me years after we intersected in New Haven.

Paul Rudolph came to his well-reasoned pedagogical position when he himself was an architecture student at Harvard, since that was the educational legacy left by Walter Gropius. Unlike Mies at IIT or Hejduk at Cooper Union, Gropius, and much later Paul Rudolph, were open to an individual student's particular design eccentricities. Always willing to criticize student work on the student's terms, Paul's encouragement had much to do with individual improvement. Any misguided students who attempted stylistic alignment with that of The Chairman soon discovered to their chagrin the fallibility of such thinking.

The mere mention of some of those who benefited from Rudolph's pedagogically flexible teaching technique reveals stylistic differences that only begin to suggest his capacity to encourage 'la difference' among his many diverse students. Tom Beeby, Alexander Cooper, Norman Foster, Peter Gluck, Allan Greenberg, Charles Gwathmey, Robert Kliment, Giovanni Pasanella, Jaquelin Robertson, Richard Rogers, David Sellers, Henry Smith-Miller, Robert A.M. Stern and Alexander Tzonis are just the tip of an iceberg. Few teachers can claim such catholicity, but Rudolph came by it naturally.

His own graduate education at Harvard was mentored by Walter Gropius who, from a pedagogical point of view, was equally democratic in producing students of widely divergent architectural skills. The mere citing of such architects as Edward Larrabee Barnes, Harry Cobb, Ulrich Franzen, John Johansen, Philip Johnson,

Joseph Passonneau and Paul Rudolph himself suggests the same kind of openness to differentiated pedagogy based on individual predispositions that Rudolph managed to coax out of his students at Yale.

While he was chairman of architecture at Harvard's Graduate School of Design, Harry Cobb exhibited the same tendencies to put aside his own formal preconceptions so as to crit GSD students on their own terms. Rudolph, Cobb and precious few others are significant role models for those who wish to teach, at least from a Socratic point of view; otherwise architects whose signature work often defines them, tend to teach by example rather than through Socratic interaction with their students.

*   *   *

In the context of a kind of perverted Shakespearian comic relief, there was a moment during my last year in New Haven that brought back distant memories from virtually two decades earlier of the circumstances that at first led me into the field of architecture.

One bleak wintry day in early 1961, I saw a portending item posted in the *Yale Daily News* that announced that Ayn Rand, whose book *The Fountainhead* had so influenced me as an adolescent, was scheduled to present her intransigent ideological and political philosophies as a part of a colloquium that very evening at the Yale Law School. Rounding up several classmates from my master's class, we trudged across the campus through inches of fresh snow to witness the proceedings. After her performance and in a moment of selective amnesia, I somehow forgot how much Ayn Rand loathed self-sacrifice. I naively introduced myself to her by saying that it was her book that influenced me to become an architect, thus changing my life forever.

With a contemptuous look she scanned my particularly un-Roark-like figure in a demeaning way and, dripping bile, replied caustically, "So what?" Of course, I should have known better than to subject myself to such biting ridicule, but at the time my naiveté overcame me, though not to the degree that I didn't recognize devastating commentary when I heard it. In retrospect, it brings to mind the well known March 29, 1976 *The New Yorker* magazine cover featuring Saul Steinberg's cartoon looking west over the Hudson River seeing objects in rapidly diminishing perspective.

*   *   *

If you were in Paul Rudolph's design studio at the school you could expect a one hour personal critique at your drawing board once a week. Normally, he only taught the master's candidate's studio, but in my bachelor's thesis year Rudolph dropped out of teaching the master's studio in that spring semester piqued by what he considered the less-than-satisfactory level of that year's class combined with a notoriously scandalous incident involving students from that class who were caught in an embarrassingly public sexual incident that was reported in *The National Enquirer*. Thus, I was fortunate insofar as he was my studio critic for both bachelor's and master's thesis classes.

Assuming that in any particular week you had completed the necessary work that he felt was the absolute minimum needed to develop your project thoroughly both spatially and materially, he would condescend to see you again the following week. Such work included, but was not limited to site plans and floor plans, sections and section perspectives, elevation studies at varying scales, construction details and both small- and large-scale chipboard study models elaborating one's conceptual and constructional concepts. If, however, you didn't produce satisfactory evidence of project development every single week that he was available to the studio to thereby satisfy his insatiable search for perfection through perpetual trial and error, you would never see him again until the final jury. By then it was too late and you could rest assured that danger would be served up on the agenda when you publicly presented your project.

Interestingly, he expressed a very real lack of interest in historical research and possible planning precedent; his focus was on the trial-and-error mechanisms that artists use to refine their work, not academic rationalizations that might otherwise give good reason for design choices by student architects. His was the standard fare of that time and was entirely different from the design studio and its raison d'etre at the turn of the 21st century. While I personally gained much by being subject to Rudolph's rigorous pedagogical methods, the best of 21st-century architecture students are, in my view, immeasurably more sophisticated in their project development. The presence of history, theory and criticism in architectural studios over time has vastly improved the student's approach to design.

*　　　*　　　*

I experienced one final bout of immense insecurity virtually at the conclusion of my final year at Yale. My master's thesis was based on a project that I had worked on while I was employed at SOM the summer before. It entailed plans for the new campus of the University of Illinois at Chicago (initially known as the Circle Campus, now referred to as UIC). One of the sites under study at the time, but subsequently abandoned in favor of a more central location nearer downtown Chicago, was situated on Northerly Island, at that time a small single runway airport known as Meigs Field. The small airport was utilized largely by corporate jet aircraft as well as small commercial commuter carriers and was the only use to which the island was put.

Drawing of the author's master's thesis project, 1961

My approach to the project was diagrammatically simplistic, such that the campus buildings, with one exception, were to be composed of structurally expressive, diagrammatically neo-Miesian rectangular parallelepipeds. These forms were conventionally distributed orthogonally over the site per unadventurous campus planning principles of the day with a single eccentric long span structure representing the university's gymnasium that was to be the centerpiece of the overall concept.

Two days before they were due, I prematurely finished my presentation drawings in anticipation of the final jury. Instead of putting my final plates safely away until they were to have been publicly presented at the master's class jury and absenting myself from the studio until they, and I, were needed, I self-importantly sat at my drawing board with the finished ink drawings conspicuously stacked for whoever wished to see them. When Rudolph sauntered by my drafting board, I audaciously offered to show him my final drawings for the project.

That naive act was one of the more self-defeating things that I accomplished in my two years in New Haven. After scanning the project in a perfunctory manner, Rudolph asked me if I wouldn't mind if he invited the rest of the master's class over to see the drawings since he felt that there were some features that he might like to discuss with them that could perhaps benefit the rest of the class generally and their own thesis presentations specifically. I unwisely agreed only to discover that he subsequently invited the bachelor's thesis class to come to my drawing board as well. It was then that I knew that I was about to be given considerable grief for both the thesis and my cavalier attitude about finishing early.

When all of my fellow students had gathered around my desk, Rudolph launched into an invective-laden tirade. He castigated me publicly, accusing me of treating both the architecture school and my project as if both came about as the result of a "blueprint reading course" as part of a vocational trade-school curriculum. He threatened to withhold my degree if I didn't make striking changes to improve the project and then stalked off leaving me, much-diminished, as well as a stunned audience of both thesis classes in his wake. I didn't have much choice but to spend the next 48 hours *en charrette* modifying the campus design suitably so as to be persuasive in my jury presentation.

When other students who were not present at my fall from grace were informed of this incident, they became terrified that they too would be subjected to The Chairman's wrath. Purely out of fear they also buckled down to work alongside the rest of us. Needless to say, everyone in both bachelor's and master's thesis studios worked around the clock for two successive days and nights. It was business as usual at the school, but there wasn't any particular joy connected with it, it was just gruelingly hard work.

Paul Rudolph had his own uniquely Machiavellian ways of instilling terror into young designers thereby causing them to expend Herculean efforts while seeking perfection. While his methods can perhaps be called into question, the extraordinary results that many of his students achieved while they were still in school and as professionals later in life were exemplary. From a pedagogical point of view, the unanswered question both then and now is, "Does the end justify the means?"

*   *   *

At the subsequent graduation ceremony in June 1961, Mr. Rudolph asked me to stay in New Haven and work full time in his office. I pointed to the curb where my wife and my toddler son were waiting in my beat-up station wagon with its tired engine idling belching gas fumes from its tail pipe, thanked him for the offer and told him that I was burned out—I couldn't stay in New Haven another moment or I would become physically ill. For better or worse, my two-year residence at the Yale School of Architecture was complete. I had no intention of elongating it any further. Leaving New Haven with Judy and JJ, I hurriedly drove home to Chicago and to my old job at SOM.

*   *   *

I didn't stay at SOM very long. When the well-known, albeit aging *enfant terrible* of Chicago architecture, Harry Weese, who had been my visiting studio critic at Yale earlier that year, offered me a position as his chief of design in his 12-person office, it sounded too good to be true. At the time, I didn't realize that Harry needed a chief of design like I needed a hole in my head, and after a scant six months of Harry's not accepting a single idea that I proposed to him and substituting sarcasm for collegial collaboration, I resigned. Clearly, my choices were now limited. It seemed to me that it was time to begin my own independent architectural practice.

Young architects chafing at the bit to open their own shop often ask me when they should strike out on their own. My answer is always the same, "When you are no longer employable, that is, when you

111

can no longer take direction, when you think you know more than your boss." Opening an independent architectural practice isn't necessarily a function of projects that you might have in hand, or money that you might have saved, or inherited, in anticipation of an architectural emancipation proclamation. Of course, one soon discovers that if you can't take direction from another architect, you will certainly have trouble taking direction from a client, but that's another story.

Architects about to strike out on their own also ask veteran practitioners like me where it is that they should open a practice. Again, my answer is always the same, "You should go home." No matter where they come from, their greatest opportunities come about at the place where they have the utmost longevity, vis-à-vis their home. No matter what side of the tracks they come from, no matter that one city or another might be more or less responsive to ideas or, for that matter, might be more or less disposed to cutting-edge architecture, the would-be practitioner can be most credible where he or she has planted the deepest roots.

I'm well aware that it seems to be an effortless statement for me to make considering that I have lived my entire life in Chicago, the most architecturally modern city on earth, but as an example of the validity of the general principle, one of my Yale classmates, Jason (Jack) de Cell who was from Yazoo City, Mississippi, went home to become one of the most influential architects in the entire Delta region. In other words, if you stick it out in one place your entire life, the chances are that you will be rewarded for your loyalty.

In any case, the worst possible thing that a young architect can do is to move geographically from place to place. If you don't go home to begin your internship, your best alternative is to go to one location and then stay there for the duration of your professional life.

* * *

My own independent practice began under the nearly quarter-century long mayoral regime of Richard J. Daley. Forty seven years later, it exists, until recently, under the equally long tenure of his eldest son, Richard M. Daley, arguably the American mayor most committed to a sustainable future, and for all intents and purposes, Chicago's 21st-century version of Daniel Burnham. If I haven't exactly "stayed the course" in terms of focusing on one particular approach to architecture at least I stayed in one place; my city of origin. I would posit that cities have powerful institutional memories, particularly when two of its mayors named Daley have held that office virtually for a half century. While the reward for loyalty may not necessarily result in the receipt of architectural commissions, that's actually beside the point. It's the precise political analogue of Umberto Eco's story of Cistercian monks, who come to understand the value of process as well as product. If we develop a sense of the importance of participating in giving back to your place of origin, that should be reward enough.

The best way that I can illustrate this maxim from a Chicagoan's point of view is to refer to Father Andrew Greeley's theme of his one act of a three-act play performed in the early 1990s at Chicago's experimental Victory Gardens Theatre. The other two acts were authored by the mystery writer Sara Paretsky and me, neither of which were frankly memorable. Father Greeley, an elderly Catholic priest is a noted author of steamy sex novels as well as a journalist weighing in on the issues of the day in a weekly column in the daily *Chicago Sun-Times* newspaper.

In brief, three actors portraying such Chicago historical figures as Ms. O'Leary owner of the infamous cow of the Chicago Fire lore, Jean Baptiste du Sable, African-American founder of Chicago, and Al Capone, notorious Chicago gangster, are found sitting curbside on a Chicago sidewalk when an actor portraying the current mayor, Richard M. Daley appears on stage. He asks them to help him in "greening" the city by planting one million trees. Raucously, they all respond with a rousing "No!" embellishing their position by referring to him in an earthy, unflattering and discourteous manner, and generally acting out against him in the most argumentative ways imaginable.

As they are about to leave the stage, Mayor Daley stretches out his arms Jesus Christ-like, and in the double entendre manner of both the Catholic Church and the Cook County Democratic Party, pleads "Do this for me!" They hesitate a moment before coming back on stage, and then they rush to return clapping their hands together while saying, "Why didn't you say so?" Moral: if you want to be a part of the game, it is sufficient for you to address the problems of your own city without expecting something in return.

*   *   *

In my own case, I can trace a direct line from my Instant City "visionary" project of 1965–66, through other descendant unbuilt projects like the floating-city proposal called Urban Matrix of 1967–68, a proposed resort complex in the Bahamas known as The Kingdom of Atlantis of 1970, a hotel scheme for George Halas and the Chicago Bears called Instant Football of 1971–72 through to the book *Visionary Chicago Architecture* that I co-edited in 2005 with the city planner William Martin.

These unbuilt projects led to a proposal that I unveiled in 2006 for a new and inclusivist city planning methodology that brought a variety of interdisciplinary participants together with planners to author a new vision for Chicago. It proposed de-accessioning low density portions of the city into urban farms in favor of realizing a more linear urban plan that, for democratic purposes, would put people in greater proximity to each other and incidentally adjacent to the environmental edges of both Lake Michigan and the Chicago River.

None of these projects were ever realized, but in aggregate they represented an increasing involvement of mine in proposals that are in sync with other historic visionary proposals relating to the evolution of Chicago. Among the more unforgettable visions that come to mind are: Frank Lloyd Wright's proposed Mile High City proposal of 1956; Reginald Malcolmson's unbuilt linear city concept of 1960; Harry Weese's seductive scheme for an airport in Lake Michigan conceived in the late 1960s and further evolved by Chicago city planners in 1970; and Santiago Calatrava's 21st-century yet-to-be-built 2,000-foot-high spiral tower.

*　　　*　　　*

Not long after the inauguration of my Instant City project, it had an unforeseen impact on at least one of its critics. Soon after its conception, the large-scale model of the proposal was featured at the New York Architectural League exhibition that accompanied the Robert A.M. Stern book/catalog entitled *40 under 40*. The model was located at the entrance to the exhibition, blocking easy access to and from the show. On the night of the opening, the noted British-born Cornell architectural educator Colin Rowe tripped as he tried negotiating his way around one side of the large obstruction.

Spying me coming out of the crowded gallery on the other side of the model, Rowe's frustration found voice as he put me in my place by irritably saying, "And three Hail Marys to you Stanley!" Having just been pilloried by his countryman Reyner Banham in an open letter to the press in London that concluded by asking "Who is Colin Rowe, anyway?" and in one of those rare moments of enlightenment granted sparingly to each of us, I responded "And three God Save the Queens to you, Colin!" Like ships passing in the night, we sailed by each other without further volleys athwart each other's bow.

*　　　*　　　*

Generally speaking, however, the imaginings young architects have about visionary large-scale work are often tempered by the realities of practice which are, more often than not, grounded in actualized small-scale commissions. So it was in my own case.

Earlier, in 1957, while I was employed at SOM on the Air Force Academy project design team, I came

The author peering through his model of
Instant City, 1966

Urban Matrix, 1968

Kingdom of Atlantis, 1970

Instant Football, 1972

Visionary Chicago Architecture book cover, 2005

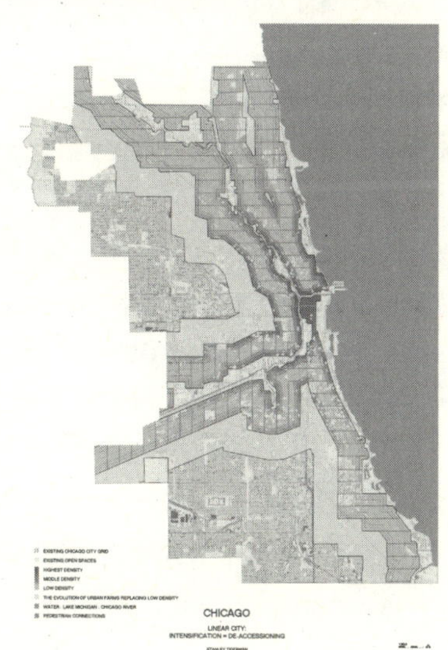

De-accessioning proposal, 2006

to be on familiar terms with a job captain working in SOM's production department on the same project. Norman Koglin and I became martini-drinking buddies, and four years later during the six months I was employed by Harry Weese, we reconnected. The two of us felt that because we represented both the design and the production sides of practice, we might make an excellent team with respect to opening an independent architectural practice. Just like that the fledgling partnership of Tigerman Koglin Architects emerged on the Chicago scene. Yet another example of instant decision making without exercising any particular due diligence, this liaison never got a real foothold as we began the slippery ascent up a peak leading to the nirvana of putting our ambitions into practice.

My mother never lived to see me begin my new life as an independent practitioner. Predeceasing my grandmother, with whom she had lived out her married life, she died in 1962 bequeathing me her life's savings of $2,000. Koglin borrowed a matching amount from his mother-in-law and the two of us rented a small, easy on the pocket two-room office at 100 West Monroe Street in downtown Chicago. We furnished the office sparsely with hollow-core doors on horses as drafting tables and secondhand stools, hired a professional appearing, well-groomed secretary and sat down to wait for a client to materialize. We didn't have long to wait.

By acting professionally, and by not trying to do things on the cheap, we gave the impression of being serious about our startup practice; another example for those just starting out. Do not start your practice in your own home. No one will take you seriously, and more to the point, they'll split hairs with you about fees, where by all rights you all should be engaged in more important discussions about the nature of the work before you.

*　　　*　　　*

But who were our clients early on, and who were those over these many decades of practice? Some of them inhabited what we designed as end users, while others were surrogates who triggered buildings to emerge on behalf of others for profit. Most were persons of some financially viable substance, though there were those who really needed what we designed for them and who stretched their limited resources as far as they could in order to realize their dreams. Most who died did so of natural causes, although two committed suicide and one came to us terminally ill.

Fortunately the vast majority of our clients were decent people, but there were more than a few who were less than admirable in their behavior to us and ours. The few that I consciously sought out tended to be generally troublesome, not paying their bills on time (if at all) and specifically predisposed to an overarching idea that precluded their being flexible. Nonetheless, the vast majority of those who chose us came because of who we were. These persons tended to be open to ideas, and by and large were quite superb clients.

*   *   *

We hadn't been in practice more than a month when my new partner's cousin, George Habenicht, commissioned us to design a startup house for him and his family near Aurora, Illinois. After the house was built, George always joked that the English translation of his German name was "have not" and, having dealt with us, that's where he ended up from a financial point of view.

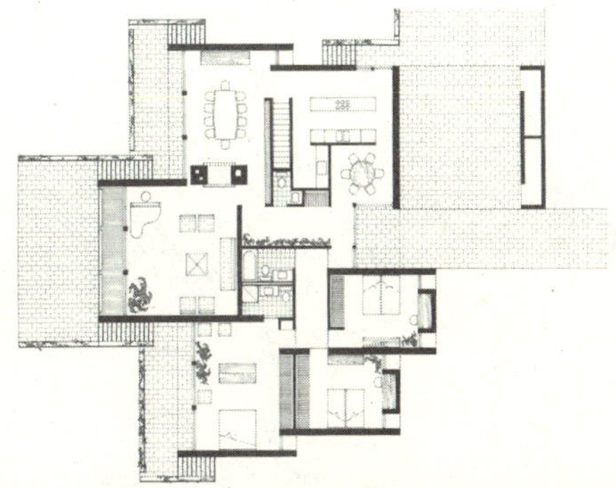

Habenicht floor plan, Elburn, Illinois, 1963

Almost immediately as well, one of my childhood acquaintances, who had visions of becoming a real estate developer, hired us to design a small three-story apartment building in nearby Evanston. With two projects in hand, we made a thoughtless decision, one that many fledgling offices often make, we worked on the two projects already on the boards without doing much of anything to obtain other new work. We were shortly left with almost nothing to keep us going.

Evanston apartment building study model, 1963

In the nick of time, a few more clients walked in the door, but not before pointless pressure caused us to begin squabbling about insignificant things that soured the atmosphere of our little practice.

"Our Ladies of the Press"

Just because our fledgling partnership began to unravel, it didn't stop me from seeking ways to expose our small practice to new clientele by means of strategically placing our work in the public realm so that people generally could gain some level of familiarity with us and with our work. While I was in the process of developing an antagonism to focused marketing to what we were capable of producing, publishing our work might have the benefit of being seen. The Zonolite advertisements that we had been commissioned to develop had an important impact on my understanding of the benefits of such exposure. I had persuaded

Rendering of one of the Zonolite ads, 1962-63

that particular corporate client, manufacturers of a masonry wall-insulation building product, that my design drawings for their advertisements could conceivably be more credible if they were presented in a manner comparable to those in the editorial section in the architectural magazines in which they were to be published.

Seeing the Zonolite ads, which featured proposed buildings published in architectural magazines, colleagues of mine began commenting about the projects that I had designed. I came to understand the value of published work precisely in the way that Paul Rudolph had understood such things before me. Soon, architectural magazines began soliciting me to publish projects that were in some stage of design development and/or under construction. More often than not, I found myself flying to New York City for the day to meet with members of the architectural and shelter magazine press for the expressed purpose of publishing my work.

As one might expect, I became friends with many of "our ladies of the press" including Jeanne (Jonny) Davern and Mildred Schmertz of *Architectural Record*, Suzanne Stephens of *Progressive Architecture*, Paige Rense of *Architectural Digest*, Elizabeth Sverbeyeff and Beverly Russell of the now defunct *House and Garden* and Susan Zevon of *House Beautiful*.

Propping up my domestic work by publishing houses that I had designed, Elizabeth Sverbeyeff almost certainly endured the most trying of circumstances in our relationship. On one occasion at a photography session Elizabeth found cockroaches scurrying around inside pots on the stove in the kitchen of a weekend house project (Hot Dog House) in a far-northwest Chicago suburb. At another photo shoot Elizabeth endured being harassed by a client who misused his advanced computer technology at her expense in his "hi-tech" aluminum-and-glass villa (Metal and Glass House) that overlooked Lake Michigan. Even so, I credit these long-suffering editors for their devotion to me early on in my career. Without their proactive support of my small-scale architectural production, I am pretty sure that the pace of my evolving career would have slowed significantly.

*　　　*　　　*

Meanwhile, our little firm had designed an eight-unit townhouse project (Pickwick Village) in Chicago's upscale Lincoln Park area for a friend of mine who had entered the Chicago real estate market as a neophyte developer. A graduate of Yale College, Donald Lebold had put together an all-Yale team of co-developers—

attorney, architect, even the sculptor (my old professor from New Haven, Robert Engman)—acquired the land and began to hype the project long before the first spade of dirt was turned.

Lebold was so intensely proud of the final product that even before the project was completely finished, when he leased the initial town home, he insisted that I see the distinctive way in which the purchaser had furnished it. Noting that there were beds and telephones in each and every pastel-painted room, and hearing that the client was an older woman with a beautiful daughter and a well-muscled son, I voiced the suspicion that things were not necessarily as they seemed. Sure enough, the weekend that the project opened for public scrutiny, a headline in that Sunday's *Chicago Tribune* glaringly printed an exposé reporting that the new townhouse owner was the madame of this establishment. She had apparently departed peremptorily from a syndicate-owned house of prostitution, her "daughter" was in fact a prostitute and her "son" was the bouncer. Apparently, her former colleagues in the mob had blown the whistle, busting her by reporting her wildcat operation to the press, and my friend, the innocent, first-time developer was thoroughly chagrined.

*   *   *

A seminal project then came our way that was to have enduring significance in the way that I understood architecture as it related to social cause. John Entenza recommended me to a controversial industrialist and art patron who had established his own family foundation that was committed to rehabilitating tired apartment buildings in some of the most rundown urban areas in Chicago. While I rehabbed numerous apartment buildings for that foundation, the most significant result of my relationship with that organization was the development of one of the country's largest low-rise, low-cost government-supported housing projects in the United States.

The 504-unit Woodlawn Gardens was the brainchild of the Maremont Foundation in conjunction with Reverend (then Bishop) Arthur Brazier, the pastor of the Apostolic Church of God on Chicago's South Side, and Leon Finney, the operating head of the newly formed The Woodlawn Organization (TWO). TWO was the antecedent organization that gave rise in some ways to Reverend Jesse Jackson's Operation PUSH. TWO also benefited from earlier community organizing by the social activist Saul Alinsky. This new community organization hired me to design this large-scale project as a corrective to Chicago Housing Authority high-rise projects.

In some ways, the development was both a benefit and a burden to a community organization that had initially come into being to resist the University of Chicago's expansionist tendencies. The community

Hot Dog House, Harvard, Illinois, 1975

Metal and Glass House, Glencoe, Illinois, 1975

Pickwick Village townhouse project, Chicago, Illinois, 1964

Woodlawn Gardens, Chicago, Illinois, 1969

felt that the university had geographically marginalized the adjacent African American neighborhood as it continued to appropriate land on the south side of the Midway Plaisance, once the site of the 1893 World's Columbian Exposition. At the time, with few exceptions, the University of Chicago was geographically situated on the north side of the Midway.

The benefit accruing to a community much in need of family housing was the medium-density, large-scale, low-rise development itself, while the burden was a lower density that did not necessarily produce the "people power" it might have otherwise had in the city council chambers had there been a greater number of dwelling units. By eschewing high-rise urban living, the new development looked to suburban masonry maisonette-like low-rise living as producing a better quality of family life that was familiar to them given the deplorable history of high-rise public housing in Chicago. The University that Alinsky had identified as the "enemy of Woodlawn's residents" continues to this very day to appropriate land for their own purposes across the Midway.

The construction of this very large project was put in the hands of the Herbert Greenwald descendant organization Metropolitan Structures. Herb's son Benet Greenwald was its project manager. Benet's former high school classmate and IIT architectural graduate Tom Hirsch represented our little firm as our project representative. At first, Tommy observed the construction process absolutely according to the procedures as defined by the contract documents and AIA general conditions holding the contractors to "plans and specs." When I suggested that he be more conciliatory, he shifted positions entirely and began consorting with contractors. I tried to make the point that construction observation is a balancing act between both extremes. Like other aspects of socially driven architecture, monitoring construction is an art that comes about through practice.

The importance of Woodlawn Gardens as a noteworthy model of the successful conjunction between social cause and architecture transcended any prestige that might have otherwise accrued to our firm. In later years, this project and others like it became paradigms for elevating my conscience level in terms of socially significant architecture.

*   *   *

Within a few months the rapport between the partners in our little practice had deteriorated to the point of divorce, and as such decrees often stipulate we divided up our projects. I kept Woodlawn Gardens, which kept me gainfully employed for the next six years. Never one to plan for the future, I fell into the habit of not "marketing." My strong feelings on that subject came about in part as a result of reading an early-1960s article in the *Wall Street Journal* about the creative director of Doyle, Dane and Bernbach, the New York City advertising agency that had popularized the Volkswagon "bug." When the reporter writing the article asked how Bernbach acquired his projects, the design partner of the advertising agency made his position on the subject clear by saying that, "If you get a job on the golf course, you can lose it on the golf course." To me, his statement not only rang true, but it also reinforced my instinctive abhorrence to marketing per se.

It seemed more important to spend the time I devoted to architecture on issues associated with design rather than wasting time on bringing work into the office. Throughout the ups and downs of fluctuating economies combined with a practice that probably could have used some periods of normalcy, I nonetheless have never regretted that decision.

*   *   *

My new, small, non-air-conditioned studio was located on the famous tree-lined North Michigan Avenue, smack in the middle of Chicago's Magnificent Mile. It was but one short block north of where I had begun my apprenticeship with George Fred Keck 14 years earlier. The office was a negligible 9 feet by 20 feet, but it was positioned on the gracefully proportioned second floor of a limestone-clad, art-deco-like, 12-story building that has unfortunately been demolished in favor of yet another expensive condominium project. Our space in this neo-classically influenced office building had high ceilings and one full-height French door that opened onto a French balcony immediately adjoining my own drawing board and directly over the front door of the building. Significantly, the top-floor tenant was the Sigmund-Freud descendant Institute for Psychoanalysis and there were numerous shrinks scattered throughout the building.

Throwing open that single French door for eight months of the year, I thoroughly enjoyed seeing one of my former kindergarten classmates, now a sober psychoanalyst (is there any other kind?), periodically coming into the building with a pricelessly somber expression on his face. Expunging his anxieties by spending the obligatory 50 minutes with his own shrink on the 12th floor, I would observe him later exiting the building sporting a relieved smile.

The studio had barely enough room for three drafting boards, stacks of bookshelves and a round conference table. We all clad ourselves in white smocks that a medical supply house renewed weekly, and the whole scene was a blast! My first employees, Larry Booth and Jim Nagle, were soon joined by Gordon

The Formal Generators of Structure, 1965-68

Crabtree, a University of Illinois, Circle Campus architecture student whose sole function in our little studio was to do research on expanding certain formal parameters that I was interested in investigating. At night after school, Crabtree documented the results impeccably in ink on vellum that was taped to a drafting board and nocturnally placed on the conference table. Labeled the Formal Generators of Structure, their subsequent publication in both art and science journals helped to cement my relationship with Chicago-based artists and local art-gallery dealers.

At their completion after innumerable months of intense labor, the Formal Generators of Structure numbered 78 impeccably drawn plates. Needless to say, in the process the student apprentice had acquired bifocal glasses. When I informed young Crabtree that, having completed our work on the subject of "orthogonal geometry" we could now enter the brave new world of "polyhedral form," he blanched, his eyes crossed, and he left my employ without so much as a "goodbye!"

*   *   *

Chicago's outsider art scene was gaining national prominence with the emergence of the Imagists, including The Hairy Who, the Windy City's own unique version of Red Grooms's inspired "vernacular-run-amok" scene in the Big Apple. Don Baum, Roger Brown, Theodore Halkin, Gladys Nilsson, Ed Paschke, H.C. Westermann and Karl Wirsum together with Roland Ginzel and Ellen Lanyon represented an artistic critical mass that had an impact on the New York art scene as well as on the Chicago architects of the day. The Chicago art galleries, some of which would exhibit architect's drawings in the future, began to be perceived as competitive with their more-established counterparts in New York and local art collectors began to patronize them.

While I was friendly with many of these artists and was a teaching colleague of Roland Ginzel, I became particularly close to Roger Brown and Ed Paschke. In 1988, Roger commissioned me to design a studio for him in La Conchita, California. Seventy five miles north of Los Angeles, La Conchita is near Port Hueneme where 36 years earlier I was in the Navy's SeaBees school. While Roger presented his thoughts lucidly in writing, his verbal skills were next to nil. Whenever we met to discuss the project I did most of the talking but virtually overnight I would receive 10 single-spaced hand-written pages from Roger outlining his reaction to my oral presentation a day earlier. I kept the entire correspondence since it represented a unique architect–artist/client relationship.

My conversations about philosophy with Ed Paschke were interrupted by his premature death, but only temporarily since he inhabits the burial plot next to Margaret's and mine in Graceland Cemetery.

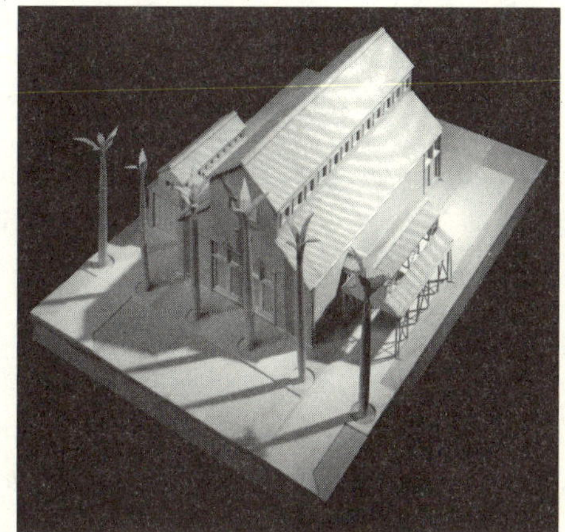

Model of the studio in La Conchita, California, 1992

In 1966 my interest in op art led me to Europe, which was funded by a grant from the Graham Foundation. I met with Bridget Riley, Victor Vasarely and other Europe-based op art painters. The trip ended in Venice where JJ and I tried fending off legions of pigeons in the Piazza San Marco.

Before the pre-eminent Chicago gallery dealer Bud Holland agreed to handle my own Mondrian-influenced, Albers-informed paintings by giving me a one-man show in the mid 1960s, he asked me a question that was to have an impact on my self-perception as an architect in relation to other architects. Bud wanted to know if I

Author and son JJ in St. Mark's Square, Venice, 1967

was interested in "flying to the moon," noting that the angle of a rising projectile was symmetrical with the angle of its descent back to earth, or were my interests more closely related to slow, steady growth. His point was that, in either case, one always "falls from grace," the only question being how much grace you take with you when, after having been in the company of angels, you plummet back to earth.

Then and there I understood that, like the myth of Icarus, flying too close to the sun can singe one's wings. The more sensible path, while staying close to the source of the flame, would be to inhabit the space just out of the spotlight's hazardously overheated beam. More than four decades have passed since the now-deceased Bud Holland made that decisive pronouncement and I am still relatively sanguine about occupying that perilous position at the edge of the architectural galaxy.

* * *

Within a week of opening the door to my new studio and before any associates joined me, I closed it provisionally for a month and left for Rome where I joined up with my former Yale teacher Paul Rudolph. The two of us then flew half-way around the world to Dacca, East Pakistan (now Bangladesh); Paul up front in first class, yours truly in steerage. My former classmate in the Yale master's program and Paul's former student, Muzharul Islam, worked in association with the World Bank to arrange for us to meet with appropriate East Bengali government officials towards the end of hopefully working on projects that were then in the planning stage in that developing country.

* * *

While we were classmates in the 1960-61 master's program at Yale, Muzharul Islam and I became close friends, in part perhaps because of another classmate, whose attitude was anything but cordial, indeed civil towards Mr. Islam. I may have misinterpreted his animosity towards Mr. Islam, but whatever the reason, I took Muz's part in a number of heated exchanges, the result of which was that Islam and I bonded. As our friendship deepened, and as many classmates before us, the two of us thought about collaborating on projects at some undetermined time in the future, never thinking that such a situation would actually ever arise.

While I returned to Chicago after graduating in 1961, Muz traveled for a year on a Fulbright Scholarship before returning to East Pakistan. Within a few years, we found ourselves teaching a design studio together for a semester at the University of Illinois in Chicago. Meanwhile, as East Pakistan's chief architect, Muz was in the process of exerting influence on his own government to hire Louis I. Kahn as the architect for what was to be known as the "second capital" in Dacca, East Pakistan. Carter Wiseman's 2007 biography of Lou Kahn has a substantial section devoted to Muzharul Islam's decisive role in that project.

As an extension of that development of the capital, and because Muzharul Islam was so keen on raising the standards of both architecture and architectural education in East Bengal, he began to lobby his government to sponsor Paul Rudolph's and my travel to his country with the expressed intent that we be employed to design projects that he felt were about to come to light. It was Muz's thought that our presence would help to raise the bar from an architectural-design point of view so as to influence young

Bengali architects towards becoming less insular and more formalistically forward thinking in their views about how to build in their own country.

In any case, Muzharul Islam was deeply interested in his country evolving a more sophisticated reputation in the development of the built environment than that which was in existence in the 1960s, and would do anything in his power to promote such a vision. Like everything else that he did and every project that he undertook, he was vocally proactive about his ambitions to better his country and lobbied anyone who might be in a position of influence to help him in achieving that goal.

Mr. Rudolph was subsequently asked by the World Bank to take on a major addition to the campus that Richard Neutra had originally designed for the Agricultural College in Mymensingh, 80 miles north of Dacca. The World Bank was then won over to commission Muz and me to work collaboratively on the design of five polytechnic institutes in five jungle villages spread throughout the province. It was the beginning of a decade's worth of unbelievable and illuminating experiences for me, unmatched either before or since that time. By delving deeply into another culture, I didn't appreciate how I would reassess my approach to architecture in the context of social cause in my own country.

*   *   *

My initial impression of one of the many manifestations of the post-colonial culture of East Pakistan came about literally the day Paul Rudolph and I arrived in that country during the monsoon season. After Muz picked us up at the airport, his driver took us to the old Shahbagh Hotel in Dacca where we checked in. Rudolph always referred to it as the "Shag Bag Hotel", which is suitably descriptive of a place that was just one step above a caravansary. Moments after we arrived at the hotel in the high midday heat and overpowering humidity of July 1964, we were quite literally accosted by a hostile British expatriate.

Clad in a dirt-encrusted, tattered bush jacket, the drunken expat appeared to be straight out of Central Casting. This unshaven, wobbly émigré had been playing snooker in the dimly lit billiards room immediately adjacent to the lobby. As we were checking in he became increasingly bad-tempered, erratically beating his bearer whenever the poor fellow brought him the wrong drink.

I was rescued from further involvement in this ugly scene when the bearer assigned to taking my bags to my quarters escorted me upstairs to my second-floor room. When he opened the door to the semi-darkened chamber, I was startled to see a furry brown spider the size of a dinner plate languishing on the threshold between my hotel room and the neighboring verandah. When I shakily asked the bearer

127

to get rid of that considerable arachnid, he calmly informed me that, "Spiders bite only Bengalis, not Americans." He then silently exited my room, leaving me alone to contend with that colossal creature. I immediately telephoned my friend Muzharul Islam to explain the situation and to ask him if I could spend the night at his place, only to hear his laughter ringing in my ear before the telephone clicked off.

*　　　*　　　*

My enculturation into East Bengal life continued the very next night when Rudolph and I had dinner at Islam's office-cum-residence with his family and friends. Typically, the male population was in conversation on one side of the dining room while the females gathered on the other. A savory buffet of local delicacies had been prepared and placed in the middle of the room, but when the meal began there seemed to be no eating utensils anywhere in sight. I noticed that everyone was eating with their bare hands, and so I followed suit. I had barely begun eating when I detected giggles emanating from the women seated on the other side of the room that appeared to be directed at me. It didn't take long before Islam laughingly told me that the custom was to eat with the right hand only since the left was used for other unmentionable and, in any case, less-hygienic tasks. I could only imagine the women's comments about the careless habits of Americans.

*　　　*　　　*

I had no idea the South Asian Subcontinent was such a political hothouse until Paul Rudolph and I attended an elaborate welcoming reception held in our honor at the governor of East Pakistan's sumptuous verandah-surrounded villa. The event was attended by an international gathering of Bengali politicians, bankers, architects, educators and officials from the World Bank and USAID. At the reception, the governor's Saville Row-tailored and Oxford-educated son confronted me and, in the King's English, blithely asked me if it was true that America's gross national product (GNP) was tied to its need to make war. Taken aback, I could only wonder at the ideological snake pit into which I had unsuspectingly been led to fend for myself.

At the same function I was introduced to a White Russian architect living in Great Britain who had studied at the Architectural Association in London before joining the World Bank as a project manager. Sergei Kadleigh was a world-weary sophisticate who was destined to be with us throughout the many years ahead that we devoted to the design and construction of those five vocational training schools. Sergei was genuinely empathic concerning a country uncertain of its future, but determined to drastically change its course. He represented an ethical approach to the future of that poor country and his proactive humanity also projected a personal image for the World Bank that, in my view, was commendable.

For many years, Sergei would travel with us to call on our construction sites that were situated in five jungle villages distributed throughout the province of East Pakistan. Ever the British colonial, whenever Sergei left Dacca for his periodic visits with us to our tropical rainforest sites, finding himself "in the bush" he only consumed his ubiquitous single malt liquor and biscuits. Each visitation represented its own matchless travel challenge. Invariably, we stayed overnight at weather-beaten guesthouses that in some distant past had been carved out of dense forest growth. These accommodations were never entirely bereft of large spiders that, after showering, somehow always ended up sitting atop my dop kit in the morning, on the face of it signifying their disinclination to share their quarters with us.

Later during that first trip we stayed in a guesthouse located in the virtually impenetrable jungle in the vicinity of Mymensingh where we had accompanied Paul Rudolph to the site of his proposed project. One night we sat relaxing after dinner on the screened verandah watching the darkened tributary of the Jamuna River slowly drift by—wondering what creatures lurked beneath its surface and good-naturedly listening to Mr. Islam incessantly railing against the foibles of capitalism. We were suddenly interrupted by the clatter coming from Mr. Rudolph's room as he banged away with his shoe at all those creepy crawlers that came out at night in the jungle. These occurrences were just the opening episodes of the chronicle of my experiences in that struggling yet seductively stunning country.

*   *   *

Immediately following that initial visit to East Pakistan, Paul Rudolph, Muzharul Islam and I traveled west together across India, first to Peshawar and then to Rawalpindi and finally to Islamabad, the capital of both Pakistan provinces located in West Pakistan, for the express purpose of meeting with officials who represented the central government. The visit was organized to formalize our working on the individual projects that the World Bank was proposing to assign us in East Bengal.

At a particular dinner party in Islamabad, I found myself in conversation with my lovely dinner partner, a well-bred, Punjabi-born, young woman with fair skin from Lahore who had just become engaged to a young man from a small village in East Pakistan. Her fiancé was also present that evening at another table across the room. She told me how distressed her parents had been at the news of her engagement and when asked why, she admitted that it was her fiancé's dark skin color that concerned her parents. Only then did I begin to understand, among other divisive issues, the depth of racial prejudice that existed between Punjabis from the desert and Bengalis from the jungle; vis-à-vis between East and West Pakistanis.

*   *   *

In 1964, when I first traveled to East Bengal as the greater region had been designated, East Pakistan was just 17 years old. It had become the province of an independent nation that resulted from the painful separation of Hindus from Muslims that was made the official Partition of India on August 14–15, 1947. By this agreement Mahatma Gandhi, Jawaharlal Nehru, Sardar Vallabhbhai Patel and other Hindu nationalists and Mohammad Ali Jinnah, leader of the All-India Muslim League, geographically separated the mostly Muslim population of East and West Pakistan from India's largely Hindu population. The assassination of Mahatma Gandhi on January 30, 1948 was only the tip of an iceberg that resulted in the ensuing communal riots as Hindus and Muslims were uprooted from their homes. Over one-million Hindu and Muslim lives were lost in that separation with more than six decades of unresolved struggles yet to come.

By the time I arrived in the South Asian Subcontinent, East Pakistan in particular had already suffered from centuries of cultural depredation. It began at the outset at the hands of avaricious Mughal emperors, was exacerbated through the heavy-handed policies of Great Britain's East India Company and then culminated in the government of West Pakistan's siphoning off much of the region's income from jute, natural gas and oil reserves in order to help finance West Pakistan's military-industrial complex.

The World Bank's support of our polytechnics project seemed to be motivated by its appreciation of the need to train Bengali supervisory personnel to help develop East Pakistan's largely untapped natural resources, thus giving the province a modicum of economic independence from its more considerable counterpart in the west. I have always admired the International Bank for Reconstruction and Development (aka the World Bank) for its sympathetic understanding of the problems facing a province trying to get its collective act together, as well as for the support that it gave to help impel important industries in East Bengal onward into the late 20th century.

*   *   *

As a young architect in my mid-30s, I saw the polytechnics project as the opportunity of a lifetime to work within a culture that had virtually no manufacturing industries to provide building material other than existing off-the-shelf rudimentary products. The alternative was to design the whole thing from scratch. Designing door-and-window hardware, sanitary fittings and millwork were among the many challenges that faced us as architects that we hoped to transform into opportunities.

The native materials, however, even when used unadventurously were of the finest quality. Burma teak was obtainable from the neighboring country to the southeast and was customarily utilized for window shuttering and other millwork. The conventional choice in East Bengal for flooring was to import marble chips and set them in a concrete matrix that would then be buffed, that is, terrazzo. Brick made from clay found along the banks of the Brahmaputra and Ganges rivers was normally used for bearing walls. Above all, the extraordinary craftsmanship of seasoned native workers in East Pakistan reassured me that whatever we designed would be carefully fabricated and would equal the skill level I had become accustomed to with Chicago construction techniques.

Our time on the project had its fill of unpredictable wacky moments. Muzharul Islam and I often traveled half way around the world to each other's offices, carrying back and forth materials and reports that were to be used on the project for research purposes. On one of Islam's visits to Chicago, for example, among his other duties he brought a single brick that had been fired at a local Bengali beehive kiln downriver from Dacca. This piece of masonry was to be analyzed by a sophisticated testing laboratory located in a Chicago suburb to determine its inherent qualities such as compressive strength per unit of measure, resistance to water penetration, salt content that might be subject to efflorescence, dimensional stability, et al.

While waiting in a glass-enclosed mezzanine for Muz to clear customs at O'Hare Airport, I observed his sartorial figure punctiliously carrying a gift-wrapped package that contained the aforementioned brick. As the US Customs official apprehensively queried him about what was inside the neatly wrapped package, my Bengali colleague would answer only that it was simply "a brick." Becoming progressively more agitated by Islam's terse and repetitious replies, and undoubtedly believing that he must have been transporting drugs or smuggling other contraband into the United States, the irritated customs officer in due course ripped the package open and smashed the brick into clay shards. Refusing to elaborate any further thus denying the authorities any satisfaction, Islam and the now deconstructed brick, its pieces safely placed inside the shopping bag generously provided by that O'Hare customs official, innocently entered the United States.

*   *   *

Before we began designing the polytechnic institutes, Muz and I persuaded the World Bank as well as our client, Dr. Waquar Ahmed of Pakistan's Directorate of Technical Education, that it would be useful for us to prepare a wide-ranging master plan for the project that might impart information that we could then break down and organize typologically, which was currently unavailable as a single document. The purpose of our master plan was to educate the up-and-coming generation of Bengali architects about such things as their own micro-climate and seismic issues. These students had just occupied their newly founded architecture school, which was a part of the Engineering University in Dacca and was supported by USAID and staffed by Texas A&M University faculty.

It was interesting to note that the Texas A&M architecture faculty assigned by USAID to bring about the new architecture school pointedly did not fraternize with any of the local architectural educators or practitioners. They seemed to favor spending much of their down time at the U.S. Consulate in Dacca. Some of us suspected that the Texans were attracted to the consulate for reasons other than the quality of the cocktails and hors d'oeuvres served there for late afternoon get-togethers.

The Texas A&M faculty designed and erected Western-style housing for their own use that rarely respected in its orientation the conventions of north–south alignment, which might otherwise address the demanding local micro-climate conditions by taking into account, for example, wind direction and sun screening. Nor did the faculty automatically provide large-scale openings in their buildings for the purpose of through and/or cross ventilation to capitalize on available breezes to naturally comfort condition their housing. Instead, their home environments were controlled artificially. However, because of local voltage irregularities electricity was not always evenly regulated, which tended to result in air-conditioning units that were only intermittently powered. Given the American's anti-social behavior towards the native population, the visitors seemed to be the ideal recipient of the Subcontinent's hostile micro-climate as a by-product of intermittent power failure.

The Americans were not the only ones who designed in ways that didn't always acknowledge the harsh circumstances of the South Asian Subcontinent, or for that matter respect the local customs specifically or the Muslim religion generally. When I arrived in 1964, the respected Greek planner-architect Constantine Doxiadis had recently completed housing in the old part of the city and I persuaded Muz to take me to see the results. Superficially, the project was compelling. Composed of alternatively disposed one-story solids and voids, the checkerboard pattern as seen from the sky flying into Dacca was textbook modern. Unfortunately, the proportion of the courtyards was such that air movement in those spaces was virtually nonexistent. Kitchens and baths were situated on the windward side of bedrooms and sitting rooms causing them to be relatively odiferous. Squatting slabs were oriented so that one's buttocks faced Mecca (not precisely respectful of the largely Muslim population!).

Interestingly, the locals living in these quarters took it upon themselves to demolish walls to more appropriately accommodate both religion and climatology. So much for the benefits of abstract architectural philosophies bereft of local religion, culture and tradition.

*   *   *

A major article on our master plan, which detailed such local conditions as microclimatology, seismology and building materials appeared in the September 1968 issue of *Architectural Record*. Though it has been amended many times since then, our master plan is still available for use in Bangladesh today. The preparation of that master plan, which occurred both in Chicago and in our joint office in Dacca, was not only at the end of the day useful to young Bengali student architects, but was invaluable to me and my American colleagues in helping us understand the critical nature of the local conditions.

Through efforts that were expended on the master plan, I developed an empathic understanding of the climatology, seismology and behavioral underpinnings of the East Bengal culture. It is doubtful that the final projects

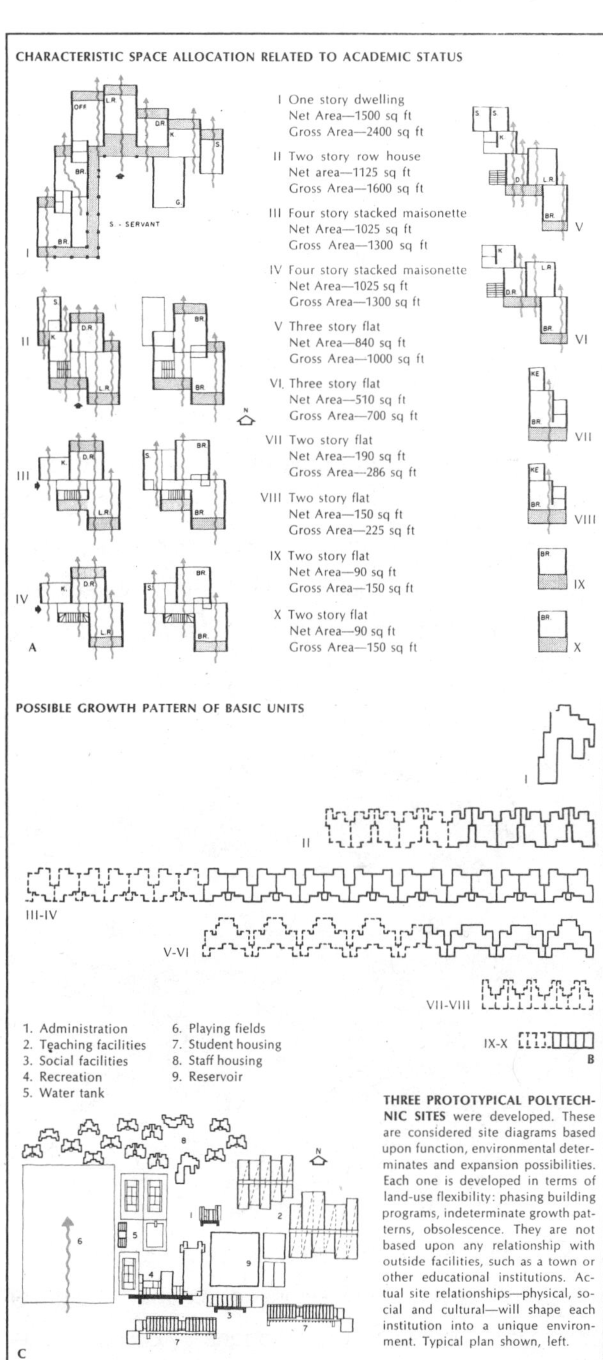

Architectural Record reprint of Bangladesh polytechnics master plan, 1968

themselves would have been as environmentally responsible as they were without our having first done the research that formed the basis of the five polytechnics master plan.

When we finally began the actual building-design process, it was clear that the structures we were to create for the vocational training institutes would not only be orientated north–south to access the maximum wind flow, but would utilize Bernoulli's principle, or the Venturi effect—the energizing of air movement to artificially increase its velocity. Wherever possible, we exploited the environmental benefits that emanated from the stack-effect theories outlined in British Overseas Weather Station Reports. In other words, the project, in detail as well as in general, needed to appropriate any and all prior methods required to contend with the extreme climate conditions in existence in the Subcontinent.

Construction methods and materials that we selected for the project included using reinforced brick bearing walls that were designed to resist seismic forces, teak louvered shutters with fly netting and terrazzo floors. Everything we did was rational and well within the modernist canon as documented in the master plan. Our architectural training as modernists eschewed the ornamental programs that we might otherwise have appropriated to symbolically represent the Bengali culture.

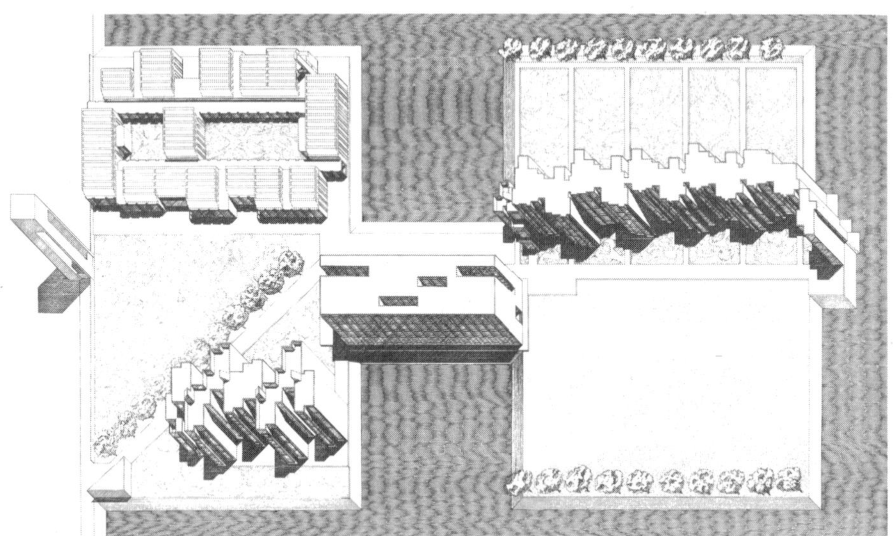

Drawing of Barisal, Bangladesh project, 1970

Four decades before sustainable issues as represented by LEED certification became *de rigueur*, we worked with local materials wherever possible to reduce the fossil-fuel emissions to transport them from abroad. The lesson that George Fred Keck taught me two decades earlier, that responsible architects need to express the way buildings are constructed, was at last beginning to yield dividends.

*   *   *

Muzharul Islam insisted that I fit comfortably into East Pakistan's cultural context so I could better understand what I was to design. To that end, we attended a variety of local social events and ceremonies, one of which included a fascinating traditional arranged wedding between two young people who had never seen each other before exchanging their marriage vows. With males in one chamber and females in the adjoining one, the newlyweds first saw each other's reversed image by peeking into strategically situated mirrors. The shock registered on their faces was priceless.

In addition, Muz persuaded me to eat the native food and to drink the local water. Almost as a matter of course, I became deathly ill with amoebic dysentery. As a result, once I recovered I became one of the few Westerners who actually gained weight while in his country.

I didn't realize at the time that I might contract amoebic dysentery again more than a half-century later. This time, I was stricken while traveling through Egypt, and with more calamitous consequences since my autoimmune response apparently is not what it once was.

As the years passed, I attained a comfort level after traveling to and from the Subcontinent on a regular basis. Given the extreme heat and the unbearably high humidity unrelieved by the most minimal wind movement (never much more than 2 miles per hour), I often dressed much as the native Bengalis did. That meant wearing loose fitting clothing, such as the lungi or punjabi and

Student housing, Bogra, Bangladesh, 1975

pyjama, as well as behaving more deliberately and moving more slowly, as opposed to the hustle and bustle associated with my faster-paced Western lifestyle.

Even so, vellum drawings taped down to drawing boards in Muz's drafting room were considerably contoured the morning after they were drawn due to the steamy tropical climate and needed to be re-taped before work could once again proceed; a small price next to the lessons I learned in that exotic country.

*   *   *

At the beginning of the 1967 monsoon season on a fine afternoon in May, I was waiting in the lobby of Dacca's Intercontinental Hotel to be picked up by Muzharul Islam's driver who was to take me to meet Muz at a government-sponsored reception. Contentedly clad in punjabi, pyjama and open sandals, I was approached by a stranger dressed in a Brooks Brothers white-linen suit, and clearly an American, though he appeared to be more like "our man in Havana" out of Central Casting. I perked up when he seemed to recognize me saying that he knew I was an architect from Chicago. He struck up a seemingly innocent conversation and seeing how I was dressed, asked me if I was on my way to a local social gathering.

All of a sudden, I understood where the conversation was heading, and I bluntly accused him of being a CIA operative. It was obvious to me that he knew that I was going to the 50th anniversary of the Russian Revolution at the Soviet Consulate and wanted me to inform on attendees whom I might recognize. It was common knowledge to those of us who had been in the country for a period of time that government representatives of the U.S.A., U.S.S.R. and China were trying to position themselves strategically within various levels of the East Pakistan government in order to gain influence as to the disposition of the country's natural resources. The operative backed off, but I was being recruited, and I knew it. However, there were even more harrowing encounters to be found in the field.

*   *   *

On route to a 1969 construction site at Bogra, 100 miles northwest of Dacca, the ancient twin-engine propeller-driven DC-3 transporting Muzharul Islam and I crash-landed in the jungle on an old pre-Second World War, Flying Tigers airstrip that was constructed of metal cleats. The DC-3's undercarriage was hopelessly damaged in the crash, rendering the plane of no use for the return trip and there was no replacement plane to be had at that time. Because I was scheduled to leave the country the following day, Muz and I had no choice but to journey back to Dacca somehow by other means. Since we had left all but a few rupees back in the city, we traveled from Bogra to Dacca in fourth-class accommodations on the local train.

While the trip was picturesque with the overflow crowds perched on the roof of the train, it was complicated by a cast on my leg that I had sprained severely while skiing in Switzerland just before I had traveled to Dacca. Stopping every few miles to pick up or deposit passengers, waiting for long periods of time at a siding for oncoming train traffic to pass, crossing the Ganges on a barge and in all cases never exceeding five miles per hour, it took us the entire night to travel back to the capital. Needless to say, I became very much aware of the infrastructure problems that existed in that country.

In addition to the progressively more dismal straits into which the country was rapidly descending, the many problems connected with the project, a challenging micro-climate, amoebic dysentery and the ever-present large-scale eight-legged creatures, my travels to the South Asian Subcontinent were strenuous to say the least. Though I sometimes traveled to East Pakistan via Japan (I was working with the Mitsubishi Corporation on plans for a floating airport at the time), I began to punctuate my westward return trips to Chicago with layovers in Paris. For me, that city represented instant re-enculturation, rehabilitation and recreation.

After the first couple of Paris layovers, I was greeted at the Hôtel Ritz in the Place Vendôme by the waiting doorman and valet, each of whom bussed me on both cheeks saying, "Bonjour Monsieur Tigerman, ça va?" That was only a prelude to my discarding my dirty bush jacket and khakis and sinking into a steaming tub with a pitcher of ice-cold martinis awaiting my disposition on the tub's ledge. If that is self-indulgence, so be it.

On one trip back to Paris, however, as East Pakistan was moving relentlessly toward its war of liberation from West Pakistan, my Pakistan International Airlines (PIA) flight was re-routed over Sri Lanka (then Ceylon) since PIA flights were forbidden to overfly India. The net result of that elongated detour was that I missed my Karachi-Cairo-Paris Air France connection, catching the next Air France flight two hours later. When we eventually landed in Cairo in the middle of the night, it was impossible not to notice that the airfield was lit up like a Christmas tree. It seems that the earlier flight that I had been scheduled on had crash-landed, killing everyone aboard. Needless to say, I engaged in some serious drinking during that less-than-restful Cairo layover.

*　　　*　　　*

One personally challenging adventure that I remember in particular was when a number of us journeyed to the southern part of East Pakistan on a trek in search of the Royal Bengal tiger. My Bengali colleagues in Dacca endlessly teased me by referring to this magnificent creature as "my namesake." This endangered species is only to be found in the jungles of the Sunderbans, a district formed by rivulets emanating from the Ganges, Brahmaputra, Jamuna, Padma and Meghna rivers, just before they empty into the Bay of Bengal. The only other living creatures at work in that overgrown part of the world southwest of Dacca and southeast of Calcutta were the lumberjacks who even now continue to threaten the fragile ecology of the Sunderbans by clear-cutting trees for use as fuel, as well as the beekeepers who still ply their trade by collecting honey from trees that makes up that dense jungle.

Ever sardonically disposed and in anticipation of the trip, Muz made sure that I came into contact with one particular beekeeper whose hair had turned white prematurely. When the beekeeper showed up one day at Muz's Elephant Road office, a translator asked the beekeeper under what circumstances his hair had turned that color. The man then revealed an 8-inch jagged scar that stretched across the nape of his neck and down one shoulder, explaining that several years earlier he was surprised and attacked by a Royal Bengal tiger while working in the Sunderbans. In the animal's characteristic attack, the beast had leapt upon the beekeeper from behind grasping the poor fellow's neck in his jaws and shaking his body vigorously in an attempt to break his neck. Failing that, the assault rendered the beekeeper comatose and the tiger dragged him back to his lair to dispense with at the animal's convenience.

When the man regained consciousness, he frantically signaled to his co-workers by blowing the whistle he kept with him for just such an emergency. His colleagues speedily appeared, beating back the tiger and affecting his rescue. A nerve-racking tale, told to me, I'm sure, so as to intimidate me with respect to my forthcoming escapade. It did the trick.

*   *   *

After flying southwest 75 miles from Dacca to Kulhna in a well-used DC-3, my wife Judy and I, together with Muz and his friend the chief conservator of forests, boarded an old wooden fishing launch in the midst of continuous monsoon rains. Live poultry and fish together with native produce that we would consume on the trip were our companions. The scene seemed a little like boarding the gangplank of a latter day, small-scale, minimum-draft, tropical-rain-forest version of Noah's Ark and just like that we embarked southward on waterways that would in due course become rivulets as they parted the jungle on their way 80 miles south to the Bay of Bengal.

Since Judy was new to the East Bengal jungle, once on board I cautioned her to relieve herself before we bunked down that first night. Narrow cushioned benches with mosquito netting were suspended from overhead lines on opposite sides of our main cabin. I cautioned her that if she didn't follow those explicit directions, when she turned on the light in the head, she would shed light on a variety of large scale creatures that would be scurrying about in the hours of darkness. Needless to say, before dawn, I awoke to hear her padding along to the facilities. Waiting with trepidation for the inevitable scream, I soon discovered that she had also dived in fright upon my mosquito netting destroying its usefulness for the rest of what became a very long night.

Some days into the trip while we were playing bridge on board our launch that was anchored midstream, bearers who had gone ashore searching for evidence of the presence of a tiger called back to us that they had discovered pugh marks that were unmistakably in evidence in the moist alluvial clay of the jungle floor. They urged us to join them to begin stalking that fearsome feline.

After we reluctantly prepared ourselves for the forthcoming ordeal, a country boat deposited us on shore. We apathetically assembled five in a line with Muz in the lead armed with an ineffective sawed-off shotgun. The chief conservator of forests strode forth behind Muz empty-handed. My wife who toted a small camera was next followed by a bearer with a machete in his shaking hand and, last in line, myself hauling a single shot .440-Magnum rifle loaded and cocked with the safety off. The significance of bringing up the rear wasn't lost on me as I had already been informed that the tiger was well-known for attacking from the rear and at a distance of no more than 25 feet from the thickened bush, a distance that the animal could negotiate in a single leap thus benefiting from the element of surprise.

*   *   *

It is worth noting that the British method of hunting the Royal Bengal tiger began by having a *machan* erected by Bengali natives in the Sunderbans. A machan is a stick-built, diagonally braced platform (not unlike a treehouse but bereft of a roof) situated aloft in a tree high enough above the ground to be secure. A live goat tethered at the base of the tree trunk would assure the hunter that the bait had been set.

When three goats had been taken away by the tiger in three successive nights, which established that a pattern was in place, the hunter was notified. The hunter then traveled from his base camp and rifle in hand, ascended the platform only to wait for the tiger to return to the scene of the crime. Needless to say, our tiger shoot was not organized that way.

*   *   *

After questionably aligning ourselves such that we were at the required distance from dense underbrush so as to draw the tiger out, we began stalking the great cat. Following its tracks perpendicularly away from the shore we saw the pugh marks turn 90 degrees to the left more or less paralleling the water's edge, then another 90 degree left turning back towards the shore. We then came to the appalling realization that we were not just tracking this creature but that it was at the same time tracking us! The pad prints of a male tiger are squarer than that of the female of the species and so we knew by the shape that our stalker was indeed a male. We nervously beat a path back towards our little country boat, promptly boarded it and rowed post-haste back to the launch where, somewhat unconcernedly we resumed our bridge game. On that trip, we never came across another Royal Bengal tiger while on land.

While on that particular trek we may not have seen our stalking feline, but that's about all we didn't see. We apprehensively observed king cobras coil around overhanging tree branches. Multi-colored birds that we heard squawking such as collared kingfishers, pheasant-tailed jacanas and swamp partridges were among the many exceptional creatures that inhabit the Sunderbans. While we carefully padded through the alluvial muck of the jungle floor, I was in point of fact relieved that we didn't come across any Royal Bengal tigers. They had become an endangered species due to illegal poaching and the clearing of forest trees by native people seeking wood for construction, at least that's my rationalization. Truth-be-told, I was scared spitless.

The only Royal Bengal tiger that we saw on that trip was seen while I stood under the canvas roof on the forward deck of our launch just aft of the prow while it plied its way through a massive monsoon rainstorm. Immediately in front of our launch we spotted a mother with her cub clamped protectively in her jaws negotiating the current while swimming from shore to shore across the approximately 50-foot width of that particular tributary of the Ganges. I have never witnessed a more beautiful spectacle of untamed maternal love raw and free in a natural setting.

*　　　*　　　*

As we often observed, everything moved at a snail's pace in the Subcontinent. Some years into the projects, while taking my new bride JoAnn on a trip to visit our construction site in Barisal, we booked the most convenient means of transportation to that village, a passage on the wildly misnamed *Rocket*. Beginning its journey southward from the Dacca port of Narayanganj as the Meghna intersected the Ganges, this paddle-wheel vessel like its land-based steam engine counterpart never traveled down the river at a speed of more than 5 miles per hour stopping every so often at jungle villages on the way to Barisal. Its so-called first-class cabins were fourth class in every imaginable sense, but the vessel had a certain cachet.

Although Barisal was less than 100 miles south of Dacca as the crow flies, the *Rocket* was clearly the most picturesque method of journeying to that particular site. Every time I traveled aboard the boat I thoroughly enjoyed myself sitting on deck and watching the opposing river banks while I listened to the cacophony emanating from the jungle creatures. They seemed to be warning us about the dangers of trespassing on their virgin turf as the *Rocket* deliberately navigated the slow current flowing south toward the Bay of Bengal.

On the other hand, the mores of the male-dominated Muslim culture could reveal themselves in unfortunate ways when foreign women unwittingly challenged their customs. One night while on board the Rocket, JoAnn and I both smoked cigarettes and imbibed alcohol publicly. I had a lapse of memory, forgetting that it is customary for Muslim women to be determinedly conscious of appropriate decorum when they are in a public place. When Muslim males observe that a female both smokes and drinks in public, it is commonly understood that her morals are questionable. We had some disagreeable moments that night and were more careful about our behavior from that point forward.

*   *   *

It was characteristic of Muzharul Islam to put the destiny of the poorest Bengali jute worker before the built environment, even though he himself was born into the East Bengal intellectual elite. Muz's father had been a professor of mathematics at Chittagong University and came from a sophisticated old East Bengal family. Muz played a significant role in the revolution that was being fomented by Sheikh Mujibur Rahman and his political colleagues. Following his return from graduate school, Muz had become deeply involved in the left-wing party politics of East Pakistan's Awami League and he had become an essential cog in the wheels that moved inexorably towards independence. In 1971 when the short but bloody liberation struggle succeeded, the newly founded independent nation was known henceforth as the People's Republic of Bangladesh.

In the spirit of honesty-in-advertising, I should report that, in addition to his pre-eminent architectural practice, Muzharul Islam was in a leadership position in East Pakistan's Marxist-Leninist political party, something incidentally that Lou Kahn could never fully appreciate. Muz felt that his country's troubles could be better mitigated primarily by his work in politics and only then secondarily through the practice of architecture.

During an utterly unplanned coincidence with Professor Kahn in the international transit lounge at Heathrow Airport in London on March 16, 1974, literally the day before he died in appalling anonymity in the men's room in New York City's Pennsylvania Station, Lou argued forcefully that Muzharul Islam's political fervency represented a pointless distraction from his commitment to his architectural practice. Hard as I tried, I failed to persuade Lou that Islam's activist involvement in the emerging struggle in his country was a considered decision. After our argumentative conversation over tea and biscuits and as my former professor and I were called to our respective flights, he reminded me that he knew that I had been a student, and was now a friend of Paul Rudolph's. He asked me that when I saw Paul next, to tell him that he very much admired his work.

Several days later, while in Dacca, I was shocked to read Professor Kahn's obituary in the local paper and shaken by the way in which he had expired. This singularly significant architect died in the men's room at Penn Station as he was preparing to return to Philadelphia following his flight from London. I cabled Paul Rudolph, making a breakfast date to meet him in the Oak Room of the Plaza Hotel on my return so I could discuss my brief visit with Lou at Heathrow Airport.

Returning from Dacca, I detoured via New York to see Paul before coming to Chicago, telling him over breakfast what I thought was a poignant tale, considering Rudolph's belligerent relationship with Kahn whenever they intersected at Yale. Instead I observed Rudolph's deadpan expression. It seems that differences of opinion about architectural theory and practice die hard, if ever.

*   *   *

After a full day's work in the drafting room of Islam's office on Elephant Road in the Paribagh section of the city, there were innumerable highly charged evenings spent sipping hot tea and arguing competing ideologies while, at the same time, trying to relax. Islam's elongated steamy verandah was adjacent to both his office and his family's living quarters. The lengthy verandah had several slow-moving ceiling fans and it overlooked his leafy, well-tended garden with its badminton court carved out of dense undergrowth. There we often listened to hot-headed young Bengali revolutionaries spiraling rants against General Ayub Khan and his central government in Islamabad as East Pakistan moved ever closer towards a struggle to liberate itself from its militarily stronger counterpart in the west. On more than one occasion, Muz had to restrain the more fanatical of these teeny-bopper student revolutionaries who wanted to travel to Islamabad with mayhem in mind, cautioning them to expend emotional capital only when they were certain that they would prevail.

In addition to my professional responsibilities during the decade that I worked there, I couldn't help being emotionally influenced by the country's unstable conditions. Its political turmoil resulted in part from the hardship and suffering inflicted on the East Bengal population by its rather militant partner that was separated from them by almost the full breadth of the Subcontinent some 1200 miles to the west.

There were numerous incidents that I observed from the sidelines that reflected the turbulence of the time. Just prior to open hostilities between East and West Pakistan and while I was at work in our office on Elephant Road, our diminutive, rotund structural engineer, Mohamed K. Shaheedullah, overloaded his miniature Morris Minor two-door sedan with thousands of bright little red-colored Mao Tse Tung-like booklets filled with inflammatory language antagonistic to the West Pakistani ruling authority. His intention was to drive to the jute fields and distribute that literature to workers. Unfortunately, just as he and his driver exited the gate bordering Islam's compound, Mr. Shaheedullah and his little red books fell into the hands of the always well-informed police who apparently had been awaiting his departure outside the compound. We later visited Mr. Shaheedullah in a top-security prison.

It all came to a head in March of 1971 when the Bengali liberation struggle erupted into a full-scale war of independence. By then, our vocational training institutes were well along in the construction process. After the war began, native Bengali construction workers who were laboring on our several sites were now and again harassed and periodically physically beaten by West Pakistani army regulars, and in at least one case, loss of life became part of the equation. When confronted by such circumstances, one either continues to work ignoring the situation, or one takes the more radical step of resigning. After too many reports of mayhem initiated by the Pakistan military on our construction workers, and after much soul searching because my country had clandestinely supported West Pakistan by providing armament in exchange for Islamabad brokering a relationship between the U.S. and China, I reluctantly exercised the latter. I did so precisely because I felt that I could no longer rationalize continuing to put people in our employ in harm's way in the process of constructing our polytechnic projects.

Having come to that difficult but, in my view, necessary conclusion, in September 1971 I traveled to Dacca during the relatively short but extraordinarily brutal war to regretfully explain my decision to our West Pakistani client Dr. Waquar Ahmed, who in turn reluctantly passed on my decision to representatives of the World Bank in Washington, D.C. I then traveled to Calcutta via Bangkok, since there were no direct flights between India and Pakistan due to the war, to tell Muz what I had done. As a known political activist, Islam had fled to India with his family immediately prior to the opening of hostilities between East and West Pakistan. He spent the duration of the war there where he had friends and relatives.

In the abbreviated time I spent in Dacca on that particular trip I was searched repeatedly at gunpoint. I couldn't help but observe the genocidal tactics of the West Pakistani soldiers who were stationed in the old quarter of the city. One evening after dinner I observed soldiers deliberately inspecting Bengali men by pulling down their lungis to ascertain their religion. If the person who was searched was uncircumcised, he was by definition Hindu and was consequently summarily executed on the spot. After the 1947 Partition, many Hindus remained in East Pakistan while most in West Pakistan were either killed or exiled to India.

Late one night, I was sequestered in the darkened apartment of our consulting electrical-engineering professor at Dacca University. Peering through a pin hole in his drawn shade I saw a number of male adults, who I was told were faculty members of the university, being lined up in the university's athletic fields by a detachment of armed Pakistani regulars. As I watched in horror, each and every teacher was shot point blank in the temple with a pistol, coup-de-grâce style.

Given these particular executions and other equally shocking displays of violent behavior, the East Pakistani perception that the government of West Pakistan was determined to reduce the eastern sector of their country once more to a solely agrarian society seemed to be borne out. I was sickened again and again as I witnessed these unconscionable acts of aggression acted out on the local population.

After arriving in Calcutta, Muzharul Islam and I scheduled a joint press conference that was held at the old Oberoi Grand Hotel to explain our decision to abandon the project. After outlining the reasons I hastily retreated to my flight back to Paris via Karachi. As we parted at the Calcutta airport, Muz's morbid sense of humor rose to the surface when he cautioned me that he was sure that the text of our press conference had in all probability already been relayed to the military authorities in West Pakistan. I was sufficiently unsure concerning the truth of that comment that when the plane landed to refuel in Karachi, I nervously remained on board anxiously awaiting the departure of the flight to the west to carry me out of harm's way.

The World Bank was displeased with my decision to not go forward with the project because of what they referred to as "political reasons," and when Bangladesh in due course gained its independence, the World Bank tried to overturn my reinstatement. However, certain well-positioned bureaucrat friends of Muz's in key positions in the fledgling People's Republic of Bangladesh government persisted and eventually I was asked by the new administration to return to complete the project. I note here that I never again received a United States government-sponsored commission.

*   *   *

As a fascinating footnote to these proceedings, when I returned to Chicago from the press conference in Calcutta, a comment was made to me by a general partner at my old architectural firm SOM who thought that I should not have resigned the commissions for the polytechnic institutes. He tried to persuade me that I would be replaced by an architect of "lesser talent." When the war ended and I was reinstated by the newly formed People's Republic of Bangladesh, this same architect suffered a convenient memory lapse when he informed me that, I must be pleased to have followed his advice. In other words, by regaining the commissions, my initial action apparently had become justified in my colleague's mind. It was a case where my determined outsider position helped me make a decision based more on the big picture at the possible sacrifice of my practice.

For that particular architect the end apparently always justified the means. I wish I could say that his views were unique to him alone, but there are too many documented cases of architects seeking an end that would indeed justify the means lest one think that this one architect was an isolated case of a less-than-laudatory approach to professional ethical behavior.

Shortly after I resigned the commissions and returned to Chicago from Dacca, the respected Bengali structural engineer and SOM general partner, Fazlur Khan, and I cobbled together an ad-hoc organization that we called the Bangladesh Defense League. Its purpose was to raise money to support the Mukti Bahini, the guerilla paramilitary organization that had been formed in the country to resist the Pakistan Army's methodical attempts to murder the country's elite. At the successful conclusion of the hostilities, we renamed the organization the Bangladesh Foundation, whose subsequent purpose became that of providing funds to the advancement of Bengali culture in all of its forms in the newly named country.

*   *   *

I returned to Bangladesh in 1974 with my second wife JoAnn to continue the work I had begun before the war for liberation. While inspecting one of our sites at Bogra, a small ceremony was arranged at the school we had designed to honor me for my having supported the new government during the liberation struggle. One of the speakers was a young Bengali polytechnic student who had fought heroically with the Mukti Bahini. One of his legs had been amputated as a result of the hostilities and yet, here he was on the podium presenting me with a memento emotionally honoring me, someone who had not done very much at all on his people's behalf.

At some level, Muzharul and Ara Islam's family became my adopted family. Their eldest son Rafique lived with JoAnn's parents in a North Shore suburb of Chicago while he attended the architecture school at

Bangladesh Defense League meeting at Fazlur Khan's Chicago apartment, 1971

the University of Illinois at Chicago. His younger sister Dahlia worked in my office on more than one occasion. We became so close that these familial relationships have endured as the Islam children became parents themselves. Before Dahlia returned to Bangladesh for good, she had settled in a Chicago suburb with her second husband and two sons thus necessitating her parent's yearly visits to a city and country subject to Muz's annual airings of his neo-Marxist rhetorical dogma. His children still call me Uncle Stan or alternatively Stanley Uncle.

*　　*　　*

After an unreasonably long hiatus of 23 years, I returned to Bangladesh in January 1997 with Margaret, Eva Maddox (the cofounder of Archeworks), and her husband Lynn. In part, the trip was occasioned by the opening of an architectural exhibition of traditional and contemporary buildings in East Bengal that featured projects designed by Muzharul Islam, which also included the five polytechnic institutes jointly realized by our two offices, the exhibition of which was mounted in the art galley of Dhaka University.

While in Dhaka, Eva and I lectured on our new school in Chicago to students enrolled in the architecture department of the Engineering University there. Later on that same trip I lectured at the architecture school at Pune, India, 90 miles southeast of Bombay (now Mumbai).

Before we left Bangladesh for India one warm overcast day we motored out of the city with Muz and his family to Dhaka's port at Narayanganj. We all then spent the next six hours on a vessel that navigated through the confluence of the Ganges, Jamuna, Meghna, Padma and Brahmaputra rivers. Dahlia's husband, Azad Nausheen, played a Bengali version of an accordion and sang native folk songs as we sailed past the brickyards and smoking stacks of kilns lining the banks on the outskirts of Dhaka until the landscape changed to small villages and jute fields.

Boats overloaded to within a few inches of the top of their gunwales with harvested products and building materials destined for Dhaka steadily plied the slow moving current upstream. One such vessel was fashioned of bamboo poles lashed together and poled through the shallow waters by several boatmen. Its destiny once it reached port was to be disassembled and then reassembled as scaffolding used to erect brick structures such as our vocational training schools.

That particular boat trip reminded me of the possibilities that might have occurred, had the confluence of these majestic, yet unpredictable rivers been subject to a major dredging program that might have turned the country into a kind of Venice of the East. Instead, during the annual monsoon rains of summer the rivers continue to overflow their low-lying banks representing the ever-present threat of death as a result. To a population desperately in need of technological answers these perennial problems remain unresolved since Partition.

Then again, not unlike the original configuration of the alluvial delta south of New Orleans, dredging might open undesirable pathways in the form of canals to the hurricanes and tidal waves that also plague that low-lying country creating even greater havoc.

More than six decades have passed since Partition and solutions to problems like the unmanageable rivers of Bangladesh remain unresolved. That abbreviated redux tour of those majestic waterways on which I had spent so much time earlier in my life brought back haunting memories of a country and its culture still very much in the process of "becoming."

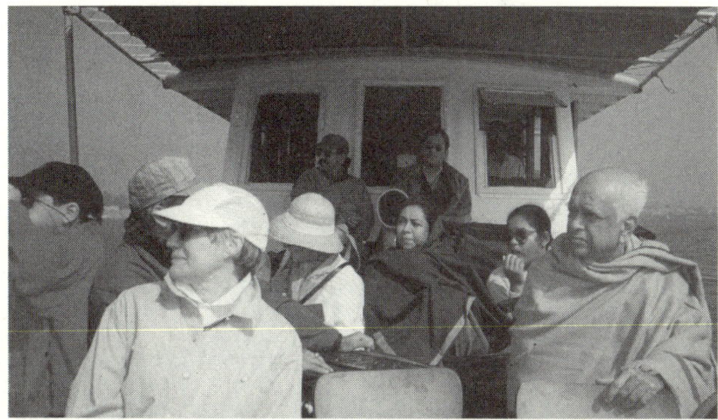

Bangladesh river cruise: Margaret McCurry (front left), Muzharul Islam (front right), Eva Maddox (in pith helmet), and Dahlia Nausheen (center)

Drawing of a Bangladesh river scene, 1997

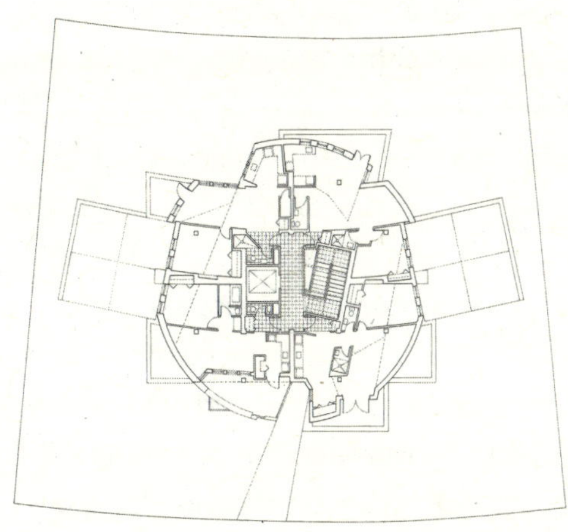

Floor plan of the Belgrade housing project, 1990

* * *

Fifteen years after the completion of the polytechnic institutes in Bangladesh, in 1990 I was commissioned to design an apartment complex in Belgrade, Yugoslavia. Not unlike both the Berlin IBA project and the Fukuoka, Japan housing complex, the Belgrade project was a part of the Third Belgrade Triennial of World Architecture and it included such architects as Justus Dahinden of Switzerland, Paolo Portoghesi of Italy, Kisho Kurokawa of Japan, Imre Makovecz of Hungary, Aleksandar Dokic of Yugoslavia, Richard England of Malta and Klaus Kada of Austria. Each of us was assigned a site atop a Belgrade hill overlooking the junction of the Danube and Sava rivers and given identical programs. Inspiration for my project came from the uneasy juxtaposition of the many nationalities that comprised Yugoslavia at the time and the resulting anxieties connected with that juxtaposition. The design for my project was as equally unresolved as the population itself.

Not long after completing the design for my project, the "troubles" that beset that distressed nation came to a head, shortly after which I resigned the commission. Though the building was eventually completed, my role in its evolution was truncated by my unwillingness to work for a government that sanctioned "man's inhumanity to man" in the form of large numbers of Muslim women who were raped during the hostilities.

* * *

During the time that I was involved in the Bangladesh polytechnic projects, I was simultaneously working on several social-housing projects in Chicago. In addition, I opened two forensic clinics staffed by African-American architectural interns in both the distressed Lawndale and Woodlawn neighborhoods of Chicago. These clinics provided pro-bono architectural services for the local population by assisting the rehabilitation of deteriorated apartment buildings for use as refurbished accommodations.

Two populist projects situated in two entirely diverse cultures halfway around the world from one another had a dramatic influence on my perceptions about the goals of architectural practice as well as a greatly increased understanding of the needs of the poor in two different cultures. Even so, it would take two more decades until I would consciously redirect my career towards a more ethical stance that would be based on my belief in the importance of doing work where poignancy needed to be embedded within a building's program to make the doing worthwhile.

* * *

Today, I find it amusing to quip that God establishes a quota for architects limiting them to a finite number of designs of up-scale suburban villas for "princes and princesses" and since I had reached that quota some years earlier, I'm no longer allowed to engage in those kinds of projects. Of course, there is some real truth attached to the irony underpinning the yarn as opposed to the current conventional wisdom that encourages specialization. Engaging in any given typology repeatedly can dull the mind and limit experimentation. An architect needs to change typologies wherever and whenever possible, at least to bring fresh insights and enriched energy to his/her architectural production thus attempting to avoid a certain *Weltschmertz*, or becoming world weary. I must admit, however, that in my own case, shifting one's career path on a somewhat regular basis was also a reflection of my conspicuously abbreviated attention span as duly noted earlier.

More to the point, however, is an endemic problem intrinsic to the design of oversized residential accommodations. Small is not simply beautiful for its own sake. "Less," as Mies van der Rohe is alleged to have said, can be construed to be "more." The extravagant expenditure of fossil-fuel energy needed to power McMansions while vast segments of the world's population have little in the way of bare necessities is becoming an unconscionable condition that is intolerable to continue to pursue. Even so and whatever the scale, one of the reasons that visionary architects continue to engage in domestic architecture is to test new design theories before they find their way into larger-scale endeavors.

The mixed blessings of professional/personal relationships

In 1963, after an initial stint as a visiting critic in the undergraduate program at the architecture school at Cornell University and a second stint a year later in the graduate program at Washington University, I accepted a tenure track position as a studio critic in architectural design at the University of Illinois at Navy Pier. I found that I enjoyed teaching. Being able to have an influence on the way that young men and women approached architecture was satisfying in ways that weren't possible in a practice.

Almost instantaneously, Leonard Currie, then dean of the College of Architecture and Art, appointed the architectural educator Donald Hanson as the new chairman of the architecture department. The former Minnesotan in turn brought the well-known Viennese architect Wilhelm Holzbauer to town as a distinguished visiting critic. Gerald McCue, who soon thereafter became the Harvard Graduate School of Design's (GSD) new dean frequently stopped in Chicago to critique Navy Pier student projects as he traveled between his professional practice in San Francisco and his soon-to-be-expanded administrative position at the GSD in Cambridge.

For reasons beyond my understanding even now four decades after the fact, there quickly developed a real chemistry between the four of us and almost as a matter of course we began discussing ways in which we could put together a loosely held global architectural consortium that our several egos idealistically referred to as "Mega I." Paul Schweikher's former partner-to-the-manor-born, the seasoned Winston Elting who also taught at the university, became a fifth collaborator. The quintet came very close to consummating the proposed professional syndicate, until I woke up one morning in a cold sweat, having realized that the last thing in the world I wanted at that time in my life was to be a part of an unmanageable collaboration. Perhaps I didn't feel that group gropes were ultimately productive to the creative process. Collaboration struck me as the beginning of "corporate" behavior in a field that was anything but, so I opted out. Though our honeymoon was short-lived, it was a genuine honeymoon nonetheless, the energy and excitement of which added much to the quality of my life at that difficult time when my practice fraught with financial woes had shrunk to three people and my personal life was unraveling.

Many years later in 1992, after an exhibition opening of my Energy Museum project at the Aedes Gallery owned by Kristin Feireiss in West Berlin, I broke bread at the Paris Bar with Danny and Nina Libeskind who had immigrated to the city initially to design the addition to the Jewish Museum. Yet again the siren of a possible alliance beckoned. This time it was in the form of a proposed partnership between Danny and me that was to have been based in Jerusalem. Once again in the cold light of day, I backed off from yet another relationship that, for me, spelled uncertainty, if not doom. I don't know what Danny and I were thinking, but when I realized that this was hopefully nothing more than cocktail conversation, I couldn't bring myself to follow through on the proposition.

It may be that I am simply not normal partner material. I also think that I wanted to preserve my outsider standing. Sharing does not easily fit into my independently minded architectural and personal lexicon. Since I never aspired to a large scale SOM-like ambitiousness, my image of an architect was always rooted in the individual, not the collective. Two earlier marriages, two failed partnerships and any number of abandoned acquaintances are testimony to my inability to sustain most long-term relationships, my personal and professional union with Margaret McCurry notwithstanding.

*   *   *

Earlier in 1988, along with a number of other architects, I was invited by Kristin Feireiss to participate in an exhibition at a *Kunsthalle* on the Kurfurstendam in West Berlin that was organized to celebrate the forthcoming millenium. Anticipating the demise of the Berlin Wall, I conceived of a project that I entitled "Der Berliner Mauer," ("The Berlin Wall"). The purpose of the project was to confront an archeology that would be more than simply a trace of what had transpired during those tense years of U.S. and U.S.S.R. confrontation of what was euphemistically called the Cold War.

I didn't anticipate that the wall would come down so quickly over the course of a few weeks in November 1989, nor did I understand the peculiar German mentality that would want to demolish not just the wall, but to

A model of the Berlin Wall project, 1989

obliterate any palimpsest of the existence of something that had so visibly divided their city for so long. Instead, my project retained the memory of the barricade by creating a linear park on both sides of the wall. A canal lined with sycamore trees paralleled each side of the divisive barrier. Cuts were made in the obstruction to signify where streets had existed before the time that Berlin was disjointed by foreign governments seeking leverage by parsing Germany into east and west sectors.

At a city scale, my Berlin project represented another attempt at introducing "cleaving" as an architectural element. The Brandenberg Gate became the nexus of an axis where East and West Berlin were oriented towards the rising and setting sun. The actual Berlin Wall, as built, was absent of color on the well-guarded east side. The presence of color on the west side was the result of considerable amounts of graffiti, more often than not placed there by uncontrollable crowds.

In my case, arriving at a space that resulted from cleaving had two interpretations. I have already digressed sufficiently about my original ambition, the exploitation of symmetry then rent asunder as my way of challenging tradition. The second rationale for rupturing an otherwise symmetrical form was to seek out the potentiality implicit in ineffability as my personal attempt to produce a sense of poetry in work that was, in my view, otherwise droll.

My sense of symbolism was furthered by the duality between Jews, who anticipate a Messiah who will come from the East, and Christians who revel in a Messiah who has, for them, already come from the East. As a Jew in Germany, any opportunity I had to design gave me license to further act out my outsider status.

*　　*　　*

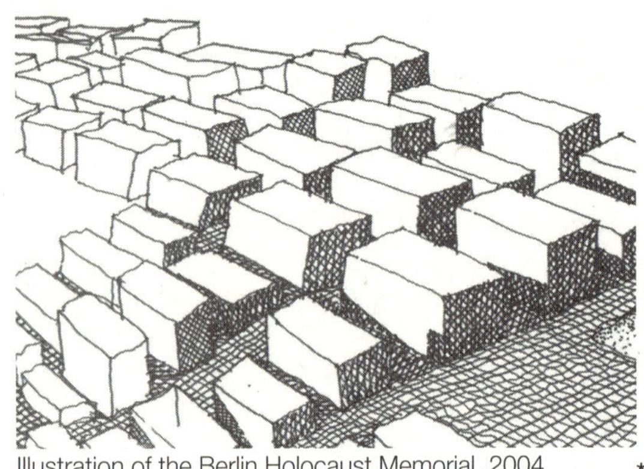
Illustration of the Berlin Holocaust Memorial, 2004

In 2004, I returned to Berlin with my wife Margaret to see Peter Eisenman's Holocaust Memorial project. The authority of Peter's use of the grid layered against the contoured archeology of the site endows the repetition of the many stele that defines the resultant *allée* with a meaning far beyond their physical presence. Those seemingly unearthed slabs are the essence of a project that doesn't succumb to representation but instead retains a sense of the *unheimlich*.

*   *   *

With respect to my teaching career, in 1971 after seven years of slogging through academic trenches at UIC, which were populated by more than a few less-than-laudatory colleagues, I sensed that I was becoming stale. Having begun to "profess" prematurely, and not being for the most part Socratic in my teaching methodology, I felt that I no longer had anything of consequence to communicate. Worse, I had developed a feeling of ennui that didn't do either my students or me any good.

Less than a year after Don Hanson left the chairmanship of the architecture school to take another teaching assignment I lost my temper during a faculty meeting held in our newly designed quarters by Walter Netsch and SOM on the city's Near West Side. In a moment of exasperation at what I considered my own as well as the administration's pedagogical ineffectiveness, I publicly resigned my tenured position as the professor of architecture. Never one to mince words, my precise comment was, "Take your tenure and shove it up your ass!" With that, I abruptly exited the faculty meeting walking away from the school for the rest of the decade. As an outsider to a faculty that found concurrence in their modus operandi, it simply wasn't suitable for me to continue in my position at the school of architecture.

But even before that fateful faculty meeting, I had become something of a pariah at the school. A controversial figure, my demanding ways with architectural students in a freshman visual-fundamentals class was not lost on the larger student body. The unrest came to a head when, after publicly tearing up certain student drawings that failed to meet the ideals that I had set for a particular class project, my own image in turn was hung in effigy outside one of the converted campus buildings that housed the design studios of the architecture school.

I did not return to that institution for nine years. For the rest of that difficult decade, my modest professional practice was no longer subsidized by teaching, except by infrequent nomadic stints as a visiting critic in architecture schools around the globe such as Harvard, Yale, Iowa State and the Bangladesh University of Engineering and Technology. Formally, my work was schizophrenic. By turns, I experimented serially with modernism, surrealism and historical allusion.

I also became a perpetual runner-up for projects for which I interviewed, and I failed to win competitions that I entered. For my proposal for the DOM office building outside Frankfurt, Germany I received an honorable mention, losing the commission to a Finnish architect. I lost the highly publicized Fargo, North Dakota bridge competition to Michael Graves as well as losing a prized Columbus, Indiana high school project to Richard Meier. The 1970s was clearly not my decade. I was a mess.

To say that my office was in the doldrums financially is a colossal understatement. The few clients for whom I labored paid slowly, if at all. One of them met his monthly obligations by paying our fee furtively in an unmarked brown paper bag in cash delivered monthly to our office by an anonymous third party. When bank loans at conventional interest rates dried up, to meet the payroll they were quickly supplanted by "juice loans" garnered from disreputable sources at usury rates. Divorced from my first wife in 1968, I had become a "weekend father" for my two children with all of the attendant emotional insecurities common to a divorced parent.

I was paying a price for insisting on doing what was necessary to insure my outsider standing. It seemed as if my 40s were ushered in as I became locked in the throes of a good old-fashioned mid-life crisis, which is not to say that certain incidents that came about didn't lighten the downside of that dismal decade.

In 1971 for example, my second wife JoAnn and I attended a celebratory reception in New York on the occasion of a newly designed apartment by Paul Rudolph. In addition to multiple light sources focused on "meadow muffin"-like seating, that night the Central Park West interior was also filled with the architectural luminaries of the day, most of whom JoAnn had already met at earlier functions. But she noticed a well-decked out personage who appeared to be holding court and wondered aloud who he was. I told her that it was my old friend from Yale, Bob Stern, and took her over to introduce them. In his own inimitable way, Bob appraisingly looked JoAnn up and down while asking her what it was that she did. Sensing that he was demeaning her and never missing a beat, JoAnn startled those present by responding as demurely as one could under the circumstances with a well-chosen phrase that rendered

Bob speechless and caused much mirth among the assembled body. She said, "…I fuck."

For some years after that encounter, Bob could not bring himself to meet JoAnn again until one fine day four years later he surprised me with a phone call saying that he was at O'Hare Airport and that he was finally prepared to confront JoAnn. I told him that he was too late, for she had just left me. In the spirit that architects play God, I had married JoAnn in the first place because I thought that love could conquer all. Of course, I was wrong. But not all was as hopeless as it appeared.

*   *   *

In 1976, the New York architect Peter Eisenman was designated by the Venice Biennale Committee to select 11 Americans to represent the United States at the first Venice Biennale of Architecture. Aside from selecting himself, Peter chose Raimund Abraham, Emilio Ambasz, John Hejduk, Craig Hodgetts, Richard Meier, Charles Moore, Cesar Pelli, Robert A.M. Stern, Robert Venturi, and yours truly. Those selections represented a broad cross section of Peter's notion of the stylistic predilections of the moment as well as theoretical differences at that year's Venice Biennale.

The Americans were to position themselves and their work *en face* against an even greater number of European architects including Carlo Aymonino, Ricardo Bofill, Gottfried Böhm, Oriol Bohigas, Alvaro Siza, Giancarlo di Carlo, Herman Hertzberger, Hans Hollein, Lucien Kroll, Aldo Rossi, Peter and Alison Smithson, Jim Stirling, Mathias Ungers and Aldo van Eyck.

A windfall payment from a recalcitrant client for work I had done previously on a large Chicago-based high-rise apartment project gave me the economic wherewithal to take my entire office as well as my two children to Venice to participate in the festivities.

At the Venice Biennale, Europeans and Americans were each given an equal number of 1-meter by 3-meter panels on which to display their work. The exhibition was held in adjoining *saloni* in the Salt Works located across the Judaica Canal from the Piazza San Marco. All participants were instructed to subdivide each heroically scaled chamber as each continent's architects individually and collectively saw fit. The Europeans parsed their brick-clad exhibition hall by using their panels as "walls" disposed by each architect polygonally one to the other such that it appeared that each exhibitor was autonomous from the rest. The result seemed somewhat analogous to a Potemkin village. All that was missing to complete that bizarre picture was a kind of visa control between each of them. To me, the way that they communally dissected their spaces so that they might appear distinct one from the other was, as I saw it, a representation of the

Dirty Postcards drawing from the Venice Biennale, 1976

Ink-on-paper drawing of Carcassonne, France,
1976

Ink-on-paper drawing of Rocamadour, France,
1976

nationalist animosities that I had always found to be extant in Europe.

On the other hand, we Americans simplistically aligned ourselves alphabetically side by side in an almost "kindergarten chats"-like display of democracy. Even at this late date after all these years, I'm still unsure whether we were being genuinely innocent or preposterously naïve.

The distinctions between American and European architects at that Venice Biennale was anything but subtle, or for that matter even affable. Those anxious distinctions reached a fever pitch at a conference hosted by the Venice Biennale Committee in a conference hall at the Lido, which was attended by all of the architects who had been invited to the Venice Biennale. There, strongly articulated and outrageously differentiated points of view devolved into a shouting match between some of the architects from each of the two continents. Aldo van Eyck and Herman Hertzberger, both of the Netherlands, lined up squarely against my ironic Dirty Postcards project which was just a bit superfluous since my offering was a spoof on American architects in the first place. Alison and Peter Smithson of Great Britain were methodically antagonistic toward what they perceived of as a lack of ideological or even intellectual content in virtually all of the projects submitted by the American architects.

It became clear to me then that Europeans in general, and European architects in particular were, more than anything else, ideologically inclined, while Americans in general, and American architects in particular were fundamentally apolitical.

*   *   *

After everyone departed the Venice Biennale my children and I took a tedious train trip across Italy from Venice to Ventimiglia and onward across the border to Monte Carlo. We spent an additional three weeks motoring along the French Riviera from Nice to St. Tropez, where we unintentionally managed to embarrass my pre-pubescent daughter Tracy and delight my 15-year-old son JJ as we aimlessly wandered along the seemingly endless topless beaches. The three of us filled the pages of our sketchbooks with everything from cathedrals to picturesque hill towns as we drove slowly northward from the Côte d'Azur and the Dordogne through Arles, Nimes and Rocamadour. We depended on my son's rudimentary high school French to negotiate our meals and lodging as we made our way up through the Loire Valley to Paris before ending the trip in London. I returned to the office with precious little work in hand to maintain my already modest practice.

On the boat from the Continent to the British Isles after weeks of coping first with the Italian and then with the French language, Tracy was so relieved when she heard a British woman say something to her in

English that she broke down and cried on hearing her native language at last. After all those weeks abroad Tracy may have felt that she would never hear English again. I duly noted all these adventures in a daily "dirty postcard" that I sent to Margaret McCurry back in Chicago.

*　　*　　*

Irrespective of the personal and professional difficulties that I had to contend with throughout the 1970s, I tried to keep a stiff upper lip by maintaining my instinctively ironic sense of humor. In 1978 for example, I had received a Graham Foundation grant to organize a three-part exhibition featuring three architects from New York, three from Chicago and three from Los Angeles. An unknown critic, one Morris Lesser, wrote a critical essay on the nine architects and their projects also authoring the corresponding catalogue that was then quickly printed and distributed.

Subsequently, while Cesar Pelli, one of the three East Coast architects included in the exhibition, and I were discussing the catalogue over coffee in New Haven while I was teaching a design studio at Yale, he asked me how I had come across the critic Morris Lesser. Cesar was mesmerized by his essay and yet he had never heard of him before reading his dissertation in the catalog. I decided to come clean and admit that Morris Lesser was actually a play on the well-known phrase that is attributed to Mies van der Rohe: "…Less is more," or "more-is-less-sir," which is to say that Morris Lesser was my own nom de plume!

At about the time that Morris Lesser arrived on the scene, several projects in my office were endowed with their own individual ironic characteristics that I couldn't resist embellishing. In 1977 I was commissioned to design a kitchen addition to an existing house for a married couple in Wilmette, Illinois who kept Kosher. The project was quickly entitled "A Kosher Kitchen for a Jewish-American Princess" and while my proposal was never brought to fruition, the exercise did reap some rewards, one in particular being a local AIA award for an un-built project.

Kosher Kitchen model, 1978

Shortly thereafter, I received a commission to design a house in Hollywood, California for a married couple who at the time were living apart from each other. They were employed by rival television networks and they felt that their disparate working hours required a kind of separation. In spite of their dysfunctional existence, they had jointly purchased a lot in the Hollywood Hills that was literally bisected by the San Andreas Fault! The charge they gave me was to design "a

Model of Hollywood Hills house, 1974

house that brought them back together." Needless to say, architects can rarely resolve those schisms. The couple in due course divorced and the project was shelved … forever.

During the process, however, I made a joint presentation to the clients in LA at which the aunt of one of the clients was present. When the aunt made several critical comments about the house, I felt a bit uneasy until I began to realize that the "aunt" was (due to a sex change) in fact the client's father. While it was beyond my grasp, it was after all, Hollywood.

*   *   *

For many reasons, in the United States the economically challenged decade of the 1970s triggered collaborations between architects/colleagues as a pragmatic way to come to grips with issues that were otherwise too monumental to respond to credibly from an individual point of view. For some of us practicing architecture in Chicago and who had ambitions to breathe new life into history, the prevailing presence of those single-minded Miesian sycophants and their apologists in the architectural press became too much to bear.

Those same ambitions fueled our feelings that we needed to find ways to loosen up what we considered to be the too-tightly held architectural reins in Chicago so that the generation of recently graduated young architects from across the country might sense a welcoming spirit not otherwise obvious in the Chicago architectural scene at that time.

Of course, our revisionist ambitions were also not very subtly self-serving; we wanted a place in the sun and the inflexibility of the Miesian hierarchy with its own hegemonic conceits precluded that. We felt the distinct need to open the window hoping that fresh Midwestern air would blow away their stuffy one-dimensional self-righteousness. We categorically interpreted Mies van der Rohe's followers as being exclusionary to us, or for that matter to anyone other than those prepared by education or by intentionality to continue in the structurally expressive mold carried on by canonically correct descendants of Mies van der Rohe.

Personally, my involvement with the Chicago Seven was a failed attempt to become an insider, but my behavior was counter-productive to growing a larger architectural firm. Many of the others in our group did enlarge their organizations; that kind of thinking, however, just didn't fit with my insistence on architecture as an individual pursuit.

It is fascinating to note that our ambitions received virtually no support from the generation of architects immediately preceding us. What was principally galling to some of us was that some of that elder age group did not themselves subscribe to Miesian orthodoxy and could have very well supported a succession not particularly committed to continuing in the Miesian mold (Bertrand Goldberg, Walter Netsch and Harry Weese). In fairness, I think that because that elder generation wasn't responsible for the generation that succeeded them, they didn't for the most part feel any great compulsion to support us.

A notable exception to the lack of a collaborative spirit in Chicago was Skidmore, Owings and Merrill generally, and its partner-in-charge-of-design Bruce J. Graham specifically. When the firm's internationally renowned structural-engineering partner Fazlur Khan died prematurely (he was 53 one when he passed away), Bruce and I buried a hatchet that we probably never should have raised over each other's heads in the first place. It was the clash of our individual egos that caused each of us to bring that weapon into play and it was the death of our colleague and mutual friend that mercifully brought our dispute to an end.

At Fazlur Khan's wake, Bruce and I reconciled tearfully falling into each other's arms. One result of our new-found détente was our collaboration on three large scale urban-design interventions that also included working with a number of other architects. That condition is still extant in Chicago today. Our cooperation found voice with the 1986 Central Area Plan designed to embellish Chicago's Central Business District and offered to the Central Area Committee for their review. Tom Beeby, Dirk Lohan and I worked alongside Bruce and his minions at SOM in preparing the document.

That in turn led to a much more ambitious association for London's King's Cross/St. Pancras railway stations. This large scale mixed-use project included David Chipperfield and Norman Foster from London, Los Angelino Frank Gehry and the respected landscape architect Philadelphian Laurie Olin.

While the London mixed-use project didn't get off the ground, it set up some ground rules that we later employed while working together on a proposal for the 1992 World's Fair that was to be held in Chicago as the centennial of the Columbian Exposition. That team included Chicago's Tom Beeby, Austin's Charles Moore, and New York's Robert A.M. Stern, all of whom helped to co-author that equally ill-fated plan.

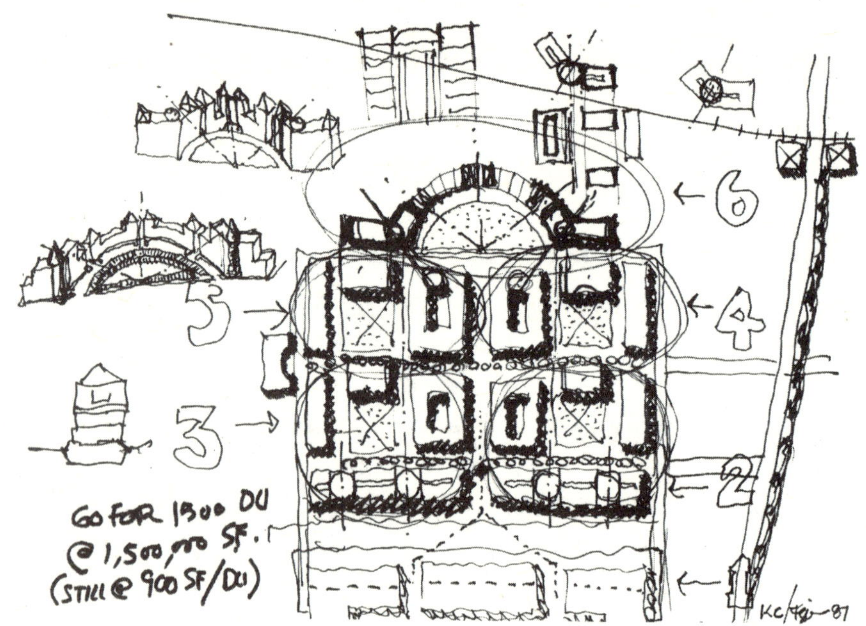

Conceptual drawing of King's Cross project/housing sector, London, UK, 1987

Conceptual drawing of 1992 World's Fair proposal, 1990

There were even charrettes that Bruce and I organized in New York, Los Angeles and Chicago that focused on a variety of proposals for a Chicago Fair. These not only included well-known architects from those three locations, but students from local architecture schools from those three regions as well.

The World's Fair project was a particularly frustrating undertaking. The proposal lost support from both the Harold Washington mayoral administration as well as the citizens of Chicago, at least in part because it was too tightly held with not even one publicly held meeting. In order for the general public to buy into proposals as vast as a World's Fair, they need to be included in the decision-making process. Contrast that with the highly successful 1893 World's Fair where virtually 2,000 public meetings were held by Daniel Burnham and the Chicago leadership at the time to debate the value of the fair as well as its legacy.

Today, that reclusive mentality is still at work with respect to the 2016 Chicago Olympic proposal. SOM was consulting on the city's ill-fated proposition with the newest generation of what is sometimes pejoratively referred to as "the luminaries:" Jeanne Gang, Doug Garofalo, Brad Lynch, John Ronan, Joe Valerio and David Woodhouse. Jeanne Gang, Doug Garofalo, Chris Lee, John Ronan and I were also jointly at work with SOM on (what turned out to be failed) proposals for the Olympic Village for the 2016 games. Again, virtually no public hearings on the process came about, unsurprisingly eliciting a similar negative response as the ill-fated 1993 rejoinder (before Chicago was eliminated early in the final round a poll was taken showing that 47 percent of the city's population favored the games while 45 percent opposed them).

*   *   *

For country-wide economic reasons alone, the difficult decade of the 1970s challenged any further expansion of "the cult of the individual" in architecture. Colleagues seemed quite happy to remove any theoretical and stylistic barriers that had separated them so as to represent a new coalescence that would make their now jointly held positions more focused and frankly less subject to attack on any one individual.

Not so in Chicago. It's not as if we had a commonly held strategy to dislodge what had come before. We already had a hegemonic authority that had held sway in our city for four continuous decades that was represented by Mies van der Rohe and his self-anointed sycophants. A number of us with precious little in common other than the yearning to open up the city to architectural dialogue were prepared to defy what we considered to be an oxygen-consuming status quo. At the time we never thought that decades later we would also be accused of holding onto the reins too tightly ourselves.

While hegemonic succession in architecture is typically generational in origin wherein "young" displaces

The Graham Foundation event, 1978

"old," in Chicago that was only part of the story. By having little in common with each other, I think a number of us knew that what we were proposing would only open the city to multivalent views, not automatically replace what we considered stagnant positions with any cohesive vision we might have bought. We didn't have a joint ideology that was shared by all. But we were alright with that.

Thus, the exhibition and catalog *Chicago Architects* of 1975-76 piloted by Larry Booth, Stuart Cohen, Ben Weese and me, followed by the ensuing founding of the Chicago Seven by which the four of us added Tom Beeby, Jim Freed and Jim Nagle. The group later bumped itself up to 11 with the addition of Gerald Horn, Helmut Jahn, Ken Schroeder and Cynthia Weese. Together we resurrected the defunct Chicago Architectural Club, first founded in 1885 by James H. Carpenter as The Chicago Architectural Sketch Club. By 1980 we had organized the Tribune Competition redux exhibition and re-issued the catalog, which revisited the 1923 original *Chicago Tribune Competition*. The newest edition included not only all of the original schemes, but also exhibited many current, albeit tongue-in-cheek proposals by architects of the most recent generation.

Revisionism in the Chicago architectural scene culminated in an architectural mosh-pit event that was hosted by Carter Manny, then director of Chicago's Graham Foundation for Advanced Studies in the Fine Arts. Tom Beeby, Larry Booth, Peter Eisenman Michael Graves, Charles Gwathmey, John Hejduk, Craig Hodgetts, Richard Meier, James Nagle, James Ingo Freed, Cesar Pelli, Robert A.M. Stern, Ben Weese and I were among those present at the well-attended event.

Gathered in Chicago to debate the issues of the day were the "whites" who were accused of being exclusivists, the "grays" who contradistinctively portrayed themselves as democratically-inclined inclusivists, the "silvers" who featured themselves as West Coast cool metal-building extrusionists, and The Chicago Seven named belatedly after their infamous revolutionary predecessors of the late 1960s.

Unfortunately, many of those who attended the two-day conference brought their own gratuitous baggage to the table; positions that engendered posturing, if not pontification. Many of those present resented others who had become more prominent than they, and there were yet others who had been students of some and had carried bitterness stemming from their lopsided student-teacher hierarchical relationships. Everyone sought some self-endowed high ground, but at the end of the day we all sensed that something unforgettable had transpired.

All of these events revolved around the fiction of a more all-encompassing Chicago architectural scene. Even so, the aggregation of those inclusivist initiatives begun in the 1970s by members of the Chicago Seven seemed in the succeeding decades to accomplish just that. Today, recent architectural graduates from

Sketch of the debate between the "whites", "grays" and "silvers" at the Graham Foundation event, 1978

around the country, as well as from around the world, come to Chicago to begin their careers seeing the city as open to a multiplicity of possibilities, not just the one that many might continue to perceive as the markedly publicized Chicago School with its structural and constructional descendant architectural population.

Of course, when you trade in direction for democracy, quality is not of necessity a predictable outcome. However, the transparency that transpired at the Graham Foundation event through debate and discussion added a real sense of potential to a newly felt Chicago architectural dialogic landscape.

I can't say that because of the circumstances that inspired the Chicago Seven to instigate events that included exhibitions, symposia and books that the initiators themselves became bosom buddies. Even so, what we did accomplish appeared to effect change, at least to a degree greater than that which had preceded us. We assumed the role that colleagues sometimes do when they are from time to time in competition with each other without any of the attendant intimacy that might have otherwise transpired had there been a greater mutuality of interest. Many of our gang affected strongly held divergent attitudes about architectural precedent and production while others had no perceivable theories to prop up their architectural preconceptions and practices. But the decade bracketing 1980 did represent a collective expression of energy that has prevailed, to some conspicuous degree, ever since.

Even though the cast of characters has changed, one of the noteworthy events that came about through the evolution of the Chicago Seven and the later Chicago Architectural Club was The Chicago Tribune Competition redux exhibition and catalogue of 1980. Beginning with the initial exhibition at Chicago's Museum of Contemporary Art, the later showings at the Walker Art Center in Minneapolis and at the Museum of Contemporary Art in La Jolla added material to the Postmodernist dialogue already in place with the movement's adherents attempting to reconnect with their architectural and cultural antecedents real or just imagined.

*   *   *

. For all of my on-again, off-again professional collaborations in Chicago, it is fair to say that aside from Margaret McCurry, there are precious few examples of any satisfying interrelationships that I can point to with any of my other Chicago colleagues. My strongest feelings of collegiality were directed towards architects who generally overflew Chicago.

Even so, a case can be made that, like oil and water, mixing colleagues with friendship introduces a certain complexity into relationships that are not always conducive to equanimity. Nonetheless, I am well aware of my often rough-edged, sometimes less-than-conciliatory personality. At some level, it can be argued that I invented the notion of "peeing on others' rugs" and I have certainly paid the price of such less than toilet-trained behavior that resulted in maintaining a small architectural practice. In any case, I have never been one to favor the political, yet alone the politically correct response to problems either of my own making or of others. Margaret McCurry's incisive comment that, "I'm now at an age where I design bridges to burn," is only transcended by her subsequent lighthearted bon mot that, "When they find you face down in an alley, it will take the investigating authorities an extraordinary amount of time to solve the case, simply because the list of suspects is so long!"

Because of the mentoring that my colleagues and I did not receive from the generation of Chicago architects who directly preceded us, such as Bertrand Goldberg, Walter Netsch and Harry Weese and virtually all the descendant Miesians with the extraordinary exception of Bruce Graham, I was determined to behave more responsibly towards the latest generation of emerging Chicago architects.

Doug Garofalo, who had been a colleague of mine at UIC and who I had first met in Rome when I

Graham Foundation "mosh pit" (from left) Helmut Jahn, Stuart Cohen, Craig Hodgetts, Michael Graves, Frank Gehry, Jim Ingo Freed, Tom Beeby, Larry Booth and Charles Jencks, 1978

Eight architects with Philip Johnson at the AIA Convention in Dallas, 1978

Press conference following the Gold Medal presentation at the AIA Convention in Dallas, 1978

was on a jury of his when he was a student in the Notre Dame Rome program, John Ronan and David Woodhouse, both of whom had worked in my office, and Jeanne Gang and Brad Lynch needed all the help that they could get in a profession not for the most part renowned for supporting its promising young architects. I think that's also why I devoted so much time and energy to the several aspects of architectural education in which I was more-or-less continuously involved paralleling my practice.

*   *   *

In my own case, the major source of mentorship, or at the very least a continuing interest in my career came from a conflicted source and not from a Chicagoan at all. Philip C. Johnson had been a visiting architectural studio critic of mine while I was a student at Yale, and like several other architects, including Peter Eisenman, Frank Gehry, Michael Graves, Charles Gwathmey, Charles Moore, Cesar Pelli and Robert A.M. Stern, I too benefited from Philip's generosity.

Who else but Philip, on the occasion of his being awarded the prestigious American Institute of Architect's Gold Medal, would commemorate the occasion by inviting the eight of us to Dallas where the annual AIA convention was located that year to lecture about our own individual work? Okay, so what if the inadvertent message sent out to all in attendance was that we were, in some way, his progeny, thus rationalizing the bestowal of that great honor on Philip.

Who else but Philip would launch the periodically held black-tie dinners at the Century Club in New York City as a modern-day salon where, of an evening, issues of mutual interest would be debated?

Certainly, many of us were well aware of Philip's checkered past. As Father Charles Coughlin's only foreign correspondent for the reactionary newspaper *Social Justice*, Philip followed the *Wehrmacht* first into Poland and then into France, reporting on the collapse of both countries in language that today would not be considered politically correct, and in any case employing phraseology that had something of an anti-Semitic ring to it. Much has been written on the subject including notably the 2008 publication *The Philip Johnson Tapes* by Robert A.M. Stern, as well as the 2009 book *Philip Johnson: The Constancy of Change* edited by Emmanuel Petit.

Century Club regulars at Philip Johnson's 90th birthday party at the Four Season's Restaurant, New York City, 1995

An unintended consequence of my relationship with Philip on the one hand, and George Fred Keck on the other was a podium design conceived by Philip for Father Coughlin, the development documents of which was authored by Keck. This dubious liaison was accomplished in the late 1930s.

William Shirer's reference to Philip's alleged relationship with the Third Reich was noted in the first edition of his *Berlin Diary* but expunged from later editions. At some level the questionable choices that Philip made while he was a young man were less than exemplary. And then there is the case of Philip's public support of Huey Long's controversial Louisiana candidacy.

From time to time I responded to Philip's early politics in ways that I'm sure he didn't appreciate nor were they my finest moments. For example, when my book *The Architecture of Exile* was nominated for a National Jewish Book Award, I showed Philip the nominating letter before one of the Century Club black-tie dinners. Understanding that I was pulling his leg, Philip asked why I did this to him. My gratuitous response was that when he came up with another candidate whom he considered more deserving, I would desist. This was clearly not the most civilized response.

On the other hand, Philip's support of our generation was singular in a profession whose history of passing the baton has been anything but exemplary. Philip presided in an ad hoc, yet forceful manner over those black-tie Century Club dinners that became legend, and he was always prepared to discuss issues that dealt with change even when he wasn't personally sympathetic to a particular topic.

Philip's openness to new ideas that did not always coincide with his own architectural predilections was acted out at one of those round-table discussions where his dissatisfaction with the official dinner topic that evening led him to change the script, asking me to explain in some detail Archeworks, the recently founded experiment in alternative architectural education in which I was newly involved.

Considering that Philip's own interest in architecture in the context of social cause was tepid at best, his proposal of the subject matter in the first place was, to say the least, unforeseen especially because it was common knowledge that Philip's reaction to the ardent Marxist architect Hannes Meyer's abbreviated tenure as Bauhaus director, after Walter Gropius and before Mies van der Rohe, was negative. After all, Philip thought of architecture as an art form, distinct from both functionalism and socialism; characteristics that Peter Eisenman would later refer to disparagingly as factors that were extrinsic to architecture.

We in Chicago had no such support from, or even dialogue with our own local predecessors who were exclusionary to the generation of architects who followed them.

The Swid Powell "stable" of architects and designers, New York City, 1985: (from left) Richard Meier, Nan Swid, Andree Putman, Arata Isozaki, Robert Siegel, the author, Addie Powell, Robert A.M. Stern, Charles Gwathmey, Frank Gehry and James Polshek

As the 20th century was coming to a close, the Century Club dinners were just one expression of a collaborative spirit that was exemplified in other circumstances as well. For instance, during the 1980s two former Knoll associates, Nan Swid and Addie Powell, started a distinctive business whose goal it was to encourage architects to engage in industrial design of household objects that Swid Powell would first manufacture and subsequently distribute. Surprisingly, the results of their combined efforts lasted more than a decade, with many of the objects becoming part of several museum collections, including notably New York's Metropolitan Museum of Art in which august institution my two china plates Pompeii and Tiger Paws reside. In November 2009, there was a major symposium at Yale about architects engaging in industrial design, of which Swid Powell played a significant role.

A substantial part of my career entailed the design of "objects," not just buildings. Beginning with flat ware for Swid Powell in New York, later with a tea service for Alessi in Italy, then jewelry initially for Cleto Munari also in Italy and later the Art Institute of Chicago and finally various small-scale objects for Jack Markuse in Boston, my architectural practice expanded well beyond the design of buildings to include decorative objects that were informed, in some measure, by use.

*   *   *

Beginning with the 1970s, the Chicago scene was, much like its national counterpart, in turmoil. Artists and architects alike were foundering in the troublesome moral and ethical uncertainties of the Vietnam era. The architectural atmosphere, however, changed slowly but surely with: the advent of the Chicago Seven in 1976; the Chicago Architectural Club in 1980; UIC's increasingly theoretical curriculum; Gene Summers' brief but noteworthy time at the helm of IIT's architecture school in the 1990s; the palpable energy at the Art Institute of Chicago under John Zukowsky's quarter-century stewardship of its architecture department coinciding with the last quarter of the 20th century; *Threshold*, the student journal at UIC, and the *Chicago Architectural Club Journal*, both published by Rizzoli; and ultimately Archeworks, commencing with its founding in 1993 and eight publications over 15 years. Slowly but surely, over the past 30 years the tide has changed and, like water seeking its own level, the descendant Miesians have finally found their place in Chicago architectural history while a new generation of evolutionarily inclined modernists has risen.

*   *   *

One of the most noteworthy efforts of the predecessor group that gave rise to the Chicago Seven was the development of an exhibition and book called *Chicago Architects*. When the 1975 exhibition previewed at the Cooper Union in New York City, it received a great deal of attention in the press in anticipation of its later Chicago opening. Peter Eisenman's Institute for Architecture and Urban Studies hosted an evening of open dialogue about the theme of the exhibition, presenting its instigators—who were in New York for the preview—as agents of some kind of *Salon des Refusés*. New York architects Philip Johnson and Ulrich Franzen mixed it up with Chicago's Bertrand Goldberg, who appeared to hold a grudge, if not outright animosity, towards Philip Johnson. It seems that the New Yorker wasn't as welcoming toward Goldberg when the Chicagoan returned from his abbreviated stint as a student at the Bauhaus as Goldberg thought he should be. Their clash was just one of several dramatic confrontations resulting from the Chicago Architects Cooper Union opening.

Chicago Architects returned to a sizable Midwestern premier opening held in the lobby of Harry Weese's Time-Life Building. When seen in contradistinction, the exclusionary *One Hundred Years of Chicago Architecture* exhibition co-curated by Franz Schulze and Peter Pran was exhibited at the same time at Chicago's Museum of Contemporary Art. The coincidence of the two disparate shows enlivened debate and discussion thereby generating an enormous amount of publicity about the hothouse Chicago architectural scene for many successive years to come.

*   *   *

My own haphazard life became somewhat stabilized with the help of architect friends who, from time to time, volunteered to come to my aid professionally, emotionally and even on occasion, financially. In 1976, the year I served as chairman of the AIA Committee on Design, I met Frank Gehry at the initial meeting of my ascendancy and we instantly became fast friends. Some months later at a subsequent COD meeting, sensing my ongoing financial plight, he offered me a substantial no-interest, long-term loan. While I declined his offer, the generosity of it touched me. On occasion, our families have vacationed

The "three heavyweights" (from left) Richard Saul Wurman, the author and Frank Gehry in Los Angeles, 1985

Peter Eisenman and the author at the Art Institute of Chicago, 1985

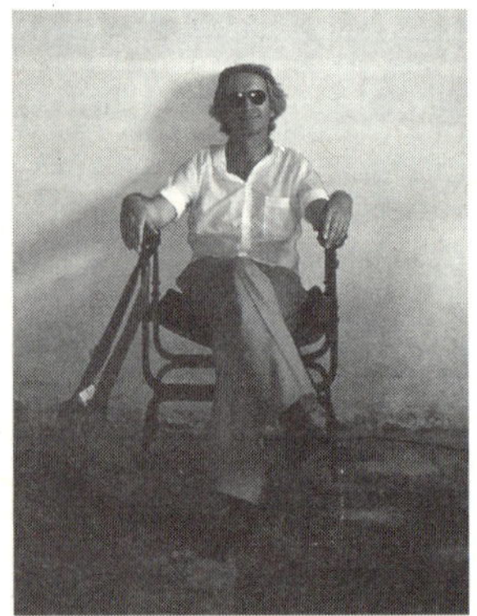

Michael Graves at the American Academy in Rome, 1980

together and years later in the spring of 1993 he pointedly came to my support at UIC in a time of my own personal crisis by lecturing at the school on my behalf.

Peers of mine came to my rescue with varying kinds of professional largesse. Peter Eisenman included me in the 1976 Venice Biennale and in the spring of 1993 he also lectured at UIC on my behalf. Eisenman also included me in the 1981 Charlottesville Tapes conference nominally entitled "P-3" that was subsequently published as a book by Rizzoli. Michael Graves recommended me for a project in Fukuoka on the south island of Kiushu in Japan and a lecture in Osaka. And in 1987, Charles Moore together with the Berlin architect Josef Kleihus invited me to participate in Moore's Tegelerhafen housing project in Berlin as a part of the IBA. The duo also asked Antoine Grumbach, John Hejduk, Paolo Portoghesi, Bob Stern and Moore to design small villa-like apartment blocks in the project. While Moore was responsible for the Tegelerhafen housing project, Kleihus was responsible for the entire IBA.

An important planning project in Europe was initiated by the Walt Disney Company's CEO Michael Eisner and Disney Board Chairman Frank Wells when they were at the helm of that organization. In addition to the Disney Imagineers who always designed the Magic Kingdom, the two executives invited well-known architects to develop planning strategies for the emerging Euro Disney project on land situated at the eastern gateway to Paris. Frank Gehry, Michael Graves, Robert A.M. Stern, Robert Venturi and I were asked to do some preliminary planning for an entertainment center whose scope encompassed the scale of a "new town."

With my inimitable tendency to put my foot in my mouth (all on behalf of perhaps retaining my outsider position), I criticized Eisner for selecting a site that I felt the French wouldn't be enthusiastic about because it was on the east side of the city on the path taken by the German army in both world wars when they entered and occupied Paris. That assessment guaranteed that beyond working on the preliminary master plan itself, I was never asked to design anything further for the Euro Disney project.

*   *   *

As part of our global network, many of us recommended each other for lectureships, teaching positions, juries and symposia; so while the 1970s may have had its downside, it was during that decade that I was also reassured that colleagues and friends were there for me even as those same colleagues independently advanced the course and the pace of their own careers.

As you might expect, I became good friends with certain of my colleagues; those with whom I shared a vision about architectural theory and/or practice and in a few cases those with whom I disagreed deeply about certain intrinsic architectural qualities. Peter Eisenman is a case in point. Peter's mantra is, much like Philip Johnson's "architecture for its own sake," that is, architecture-qua-architecture. On the other hand, I feel that architecture is not merely aesthetically driven but is also a social art that is best measured by significant others. Nonetheless, Peter and I have coexisted for 35 years, even if he did boycott my wedding to Margaret McCurry. (A Mafia-like breaking-a-finger move to punish me for attending MOMA's opening that excluded John Hejduk from a show that the rest of us were included in.) Perhaps the two of us can best be described as being each other's dysfunctional sibling. Our conversations begin by arguing about who is more of an outsider than the other.

Frank Gehry is another friend who for over three decades has more than proven that loyalty is not an empty word. His emotional support throughout the decades continues apace, as each of us moves towards our four-score scorecard.

Without John Hejduk's 35-year friendship, I wouldn't have been tested as keenly as I was in the context of what Hejduk referred to mysteriously as "aura." For more than three and a half decades we talked weekly. Our conversations ranged widely about his frustrations as dean of Cooper Union's architecture school, his fits and starts over his work and our shared views about the state of the art of architecture. At some level, John was my poetic conscience and his premature death in 2000 robbed those of us closest to him of arguably the most poetically inclined architect at work in the United States in the last quarter of the 20th century.

The former Art Institute of Chicago's chief curator of architecture John Zukowsky not only repeatedly commissioned me to design exhibition installations at the museum, but he also recommended me on a regular basis for similar assignments to other art institutions around the world. His unprecedented quarter-

John Zukowsky and the author at the opening of the Chicago Exhibition, 1987

Chicago Architecture and Design 1923–1993 exhibition, 1993

Eva Maddox, 1998

Entrance to the Chicago Exhibition, 1987

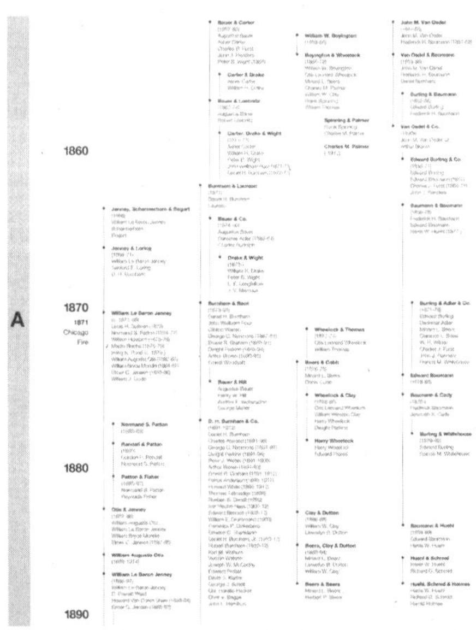

A portion of the Chicago architectural genealogy, 1993

century tenure as the chief curator of architecture at the Art Institute of Chicago represented a golden era in Chicago architecture. He supervised the recording of the oral histories of the city's major living architects and resurrected its former masters in such a way as to enhance Chicagoan's understanding of their architects and the work that extended their legacy. No Chicago institution had ever before celebrated them in quite the same way or in as great detail without conveying the notion that he was trying to bring culture to the local museum-going population.

John masterminded two blockbuster exhibitions that defined Chicago's architectural history from the Chicago Fire through the Columbian Exposition Centennial. In addition to planning and designing both exhibition installations, I authored an article in the latter exhibition catalog/book on the genealogy of Chicago architects, a document that I felt was sorely lacking in Chicago's otherwise luminous architectural library. The genealogy is an important record of Chicago's architectural continuum, albeit that it requires constant updating to assure its continuing value. In addition to an architectural "family tree," the text accompanying the chart delineates its value for purposes of linkage, influence and legacy.

With the understandable exception of my wife and partner, Margaret McCurry, the Chicago design colleague with whom I remain closest is Eva Maddox, which in all fairness has been a continuously argumentative relationship as much as it is mutually supportive. Our professional association evolved into our jointly fabricating Archeworks out of whole cloth. We threw ourselves wholeheartedly into our 15-year adventure which has resulted in certain multi-disciplinary pedagogical innovations that helped to inform the way in which the young are exposed to, and then educated in design.

The multidisciplinary aspect of Archeworks alone in combination with its being situated in a post-disciplinary setting has produced unanticipated benefits in projects undertaken and then delivered. Teaming has also reaped rewards, perhaps proving that Buckminster Fuller's notion that one plus one could perhaps produce something greater than the sum of its parts, which is to say "synergy." Eva's and my disparate design education and practice and bouncing ideas off of each other gave Archeworks a kind of a resonance and vitality that would not, I suspect, have otherwise occurred.

In 2008, a symposium was held at Archeworks entitled "Passing the Baton" which not only introduced Eva's and my successors, Sarah Dunn and Martin Felsen of Urban Lab, but also presented a cast of characters representing new design leadership in Chicago in virtually all of the institutions devoted to architecture and design. Moderated by Ned Cramer, the editor of *Architect* magazine, the issues of the day were debated by Lee Bey, the director of The Central Area Committee, Greg Dreicer, the chief curator of the Chicago Architecture Foundation, Sarah Dunn, Zurich Esposito, the executive director of the Chicago Chapter AIA, Martin Felsen, Sarah Herda, the new director of The Graham Foundation, recently appointed Hennie Reynders, the director of the school of interior architecture at the Art Institute of Chicago, Joe Rosa the curator of architecture and design at the Art Institute, Zoe Ryan, curator of design at the same institution, and Bob Somol, the new director of the architecture school at UIC.

The obvious challenge placed upon the shoulders of these next-generation leaders is synergistic in nature; perhaps demonstrating what might be accomplished through collaboration. While the evening's events didn't produce much in the way of earth-shattering visions, the existential fact of the dialogs established a starting point by which these 10 might find common ground in the future.

*   *   *

In the mid 1970s, as the result of a decanal search undertaken at IIT, the institution picked an exceptionally talented alumnus in the person of James Ingo Freed as the new dean of its architecture school. This man who we all knew as "Ingo" had graduated from IIT in 1953, and in the intervening two decades had lived in New York City ultimately becoming a partner of I.M. Pei's and Henry Cobb's. Life in that cosmopolitan city added a layer of finishing-school-like sophistication to Ingo's persona that, at the time, simply wouldn't have been possible had he stayed in Chicago. Thus, when he returned in his svelte navy blue pin-stripe double breasted suits to guide IIT's architecture school in new and unpredictable directions not hitherto known to its relatively myopic faculty, Ingo added a dimension that both benefited and, because of IIT's institutional and individual faculty truculence, ultimately challenged his position at the school.

In all events, upon his return to Chicago, the two of us renewed a friendship that began in the early 1950s when he was an architecture student at IIT. When we founded the Chicago Seven in 1975, I asked him to become a part of our group. He joined us enthusiastically, meeting with us regularly and participating in our art gallery exhibitions, our publications and most importantly engaging in much needed exchange of ideas between us so as to help clarify our individual positions, such as they were.

Before his frustration with the doctrinaire faculty at IIT caused him to return to New York for good in 1978, he was a major source of encouragement for me personally, stiffening my backbone when I produced the "Titanic" photo collage. This ironic commentary on the Miesian descendant's determination to establish hegemonic authority in Chicago showed Crown Hall (the architecture school at IIT and Mies's crowning achievement at that institution) ostensibly sinking beneath the waves of Lake Michigan.

While he remained in Chicago, Ingo was a wellspring of sophisticated interpretations of the Chicago scene and, for me personally, a loyal friend in a city that provided more acquaintances than significant friendships.

Regrettably, certain diehard factions of the IIT faculty weren't entirely enthusiastic about Ingo's attempts to unfetter a curriculum

Photo collage of "The Titanic" (Crown Hall Sinking), 1978

that still remained in lockstep with the one that Mies van der Rohe, Ludwig Hilbersheimer and Walter Peterhans left as their misconstrued legacy. In 1979 it didn't reassure the faculty when Ingo brought outside lecturers to the school like John Hejduk, the poetically inclined dean of Cooper Union's architecture school and Colin Rowe, the internationally respected historian, theorist and architecture critic at Cornell University. Ingo proposed introducing "color" into the school's core curricula thus (amazingly) galvanizing the faculty against him. The dogmatic tenured faculty became apoplectic at these proposed changes and so, after a much-too-brief three-year period as dean he returned to his partnership in New York City, leaving Mies van der Rohe's educational legacy at IIT to continue its uninterrupted descent into history with its stultified image—let loose from progressive thinking—intact.

Given that the loyalty and support shown me by many friends and colleagues over a number of years appear to prove the case that "no man is an island," my sometimes fickle nature has not been my most endearing characteristic. I would like to rationalize my way out of that reputation by holding up my short attention span as an excuse, but regrettably honesty suggests otherwise. It would also be a relief to blame others for this character flaw but finger pointing, particularly in my mother's direction, would be a cop-out of heroic proportions.

*   *   *

On St. Patrick's Day in 1979, Margaret McCurry and I were married and that very act proved to be pivotal in my own personal life in more ways than the obvious one. While the marriage itself, albeit to an architect, didn't immediately have a perceivable influence on my career, it represented the beginning of a series of events, the aggregation of which did have an important impact on my perceptions of the world around me generally, and about architectural practice/education and theory specifically.

Working in proximity to each other virtually from the beginning, our symbiotic relationship had an impact on the way that the two of us approached design independently. My normal rush to judgment attitude toward design was markedly slowed simply by Margaret's being there, while Margaret's more measured approach was made slightly more spontaneous by my hectoring.

On the other hand, we each still retain the individual dichotomous theoretical and practical architectural positions that we each brought as dowries to our liaison. It is significant that precisely because of our different approach to most things architectural, that the way in which I approach the conflictual aspects of my response to the built environment are reinforced.

*　　*　　*

During this period, my practice depended, in some measure, on lectures, juries and stints as a visiting critic for whatever minimal financial stability I was able to eke out. Even our brief European honeymoon was a lecture tour. This pattern was symbolized in some way by an invitation to participate in a seminal event that was sponsored by the Miami Architectural Club. The 1979 event was held in Boca Raton, Florida in the summer and was attended by well over 1,500 persons.

From left: the author, Richard Meier, Charley Gwathmey, and Peter Eisenman, 1979

During the three-day long weekend that we all gathered in that steamy part of the world, there was plenty of time for tennis to be played by those who thought that they were so endowed. In particular, Peter Eisenman, Charles Gwathmey, Richard Meier and I played numerous sets of doubles. Our individual conduct on the court was more or less consonant with the way that each of us approached architecture—perhaps life itself. Peter Eisenman dressed counter-intuitively entirely in black and played what can only be referred to as a "devious game." Charley Gwathmey lived up to his muscular reputation as a jock by hitting the ball as hard as he could at his opponent's genitals whenever that person was situated at the net. Richard Meier, who was impeccably clad in polo whites, played an equally impeccable and well coached game. I used my mouth a lot along with a wicked undercut.

*　　*　　*

By 1980, my architectural practice had been reduced to an office of four with scarcely enough work to sustain us, when Bill Lacy, then the president of the American Academy in Rome, asked me to spend the next three months of my life as the resident architect at that illustrious institution.

Never hesitating for a moment, and leaving my three other associates alone in the office in Chicago to fend for themselves, I hastily accepted, thirsting for the kind of interdisciplinary structure that I hoped I would discover in McKim's classical palazzo anchored atop the Janiculum Hill above Trastevere on the other side of the Tiber River.

I was to find that not only was the intellectual life at the academy splendidly stimulating in the broadest possible interdisciplinary sense, but it also gave Margaret and me a three month honeymoon that we could not have otherwise afforded.

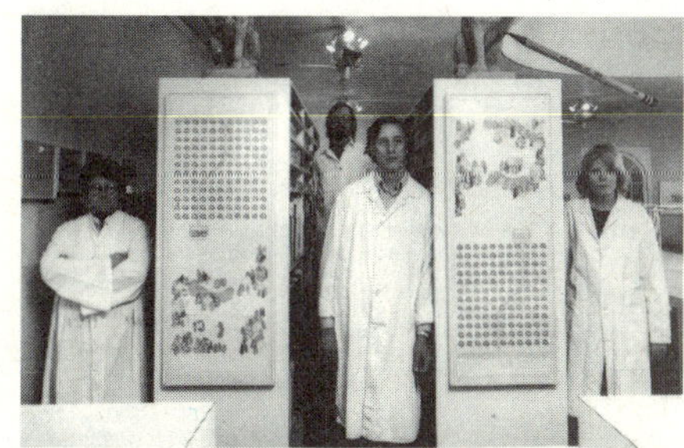

The author, Michael Abbot, Robert Fugman and Deborah Doyle in the office at 920 North Michigan Avenue, Chicago, Illinois, 1980

*   *   *

Our first morning in Rome was complicated by what sounded like a disconcerting phone call. We had been assigned the three-bedroom "Mezzanino" suite in the very apartment that Charles Follen McKim had designed for himself at the outset of the 20th century. Our quarters overlooked the *cortile* on one side and the large rear garden with its tennis court bounded by the Aurelian Wall on the other. As I soaked in a steaming tub in the large

JJ Tigerman and the author in front of the American Academy in Rome, 1980

bathroom decompressing from our transatlantic voyage, Margaret burst in, tremulously reporting that some "Mafioso-sounding" character had called and had seemed threatening on the phone, only to then peremptorily hang up. I had failed to warn her that the caller was without doubt acting out one of the numerous fictional characters that none other than Peter Eisenman employed from time to time to amuse himself, if not of necessity others.

*   *   *

Irrespective of that disruptive opening moment, Margaret and I began an idyllic period in our newly married life in Rome living atop that hill in an institution that, while it was protected by a guard with his inevitable Uzi slung over his shoulder, was defined by multidisciplinarity. The intellectual and social ambiance was enlivened by the likes of John D'Arms, the director who was on leave from his post as distinguished scholar in the classics at the University of Michigan; James Ackerman, Harvard's highly

The class of 1980 at the American Academy in Rome

The author, James Ackerman and JJ Tigerman in the back garden of the American Academy in Rome, 1980

regarded Palladian scholarly academic; Irving Lavin, Princeton's Institute for Advanced Studies Bernini scholar; Charles Simmons, *The New York Times Book Review* editor and author; Morley Baer, the west coast architectural photographer; and Bill Turnbull, the San Francisco architect. This auspicious group produced, at the very least, a dazzling, intellectually informed atmosphere. Scholars, practicing artists and architects, historians, writers, musicians and intellects of all kinds combined comfortably in that *Last Year at Marienbad*-like sanctuary as they intersected with each other both in a structured as well as in an ad-hoc way.

I spent many satisfying hours in the academy's well-endowed library, played piano duets with John D'Arms, was entertained by Professor Lavin's Machiavellian scholarship methods, listened raptly to Charles Simmons speak in whole sentences and was suitably sympathetic to Billy Turnbull's turbulent tales of lost loves, as he curled up for many nights on the couch in our living room. Turnbull punctuated his stay at the academy with trips back to San Francisco, from whence he returned with bottles of his winery's rich Cabernet Sauvignon that we savored at the numerous cocktail parties that graced our apartment. Margaret and I played tennis on sunny afternoons on the academy's clay tennis court located just below our window, while at night we indulged ourselves after dinner adjacent to the jasmine-scented cortile with a game of cowboy billiards, a diversion that had been invented at the academy many decades earlier.

Sunday mornings were unique insofar as we picked our way down Janiculum Hill through minefields of used hypodermic needles to the weekly Porta Portese Flea Market. We whiled away Sunday afternoon playing volleyball in the back garden while listening to musicians playing in their studios embedded in the Auralian Wall at the rear of the property. It was a time like no other and we soaked it up.

At best, my residency in Rome was inspirational. My time at the academy led directly to writing my first book *Versus*, the theatrical design of my pavilion at the 1980 Venice Biennale (the second one in which I was asked to participate), and countless other intellectually stimulating events that had a strong impact on my perceptions about art and architecture; perhaps too many for one who cherished his independent position within architecture.

Irving Lavin inspired me to write *The Architecture of Exile*. The academy's former director of classical studies Mason Hammond, the crusty New Englander and distinguished Harvard scholar, tried to persuade me that his ex-marine, ex-son-in-law and ex-SOM general partner Bill Drake was possibly not the best choice to teach ethics in a graduate electives seminar that I was coordinating at UIC. Though I respectfully disagreed with Mason at the time, in the long run he might well have been right.

*   *   *

Residency at the American Academy offered some unexpected insights into the strikingly different ways that artists and architects approached their work. Every morning, for example, after consuming more than a few cappuccinos and cornetti in the wood-paneled bar that was backed by cartoons of all prior Rome Prize winners and neighboring the jasmine-scented *cortile*, the painters and sculptors at the academy eagerly retired to their respective studios, proactively looking forward to the day and to initiating work.

On the other hand, the architects in residence reluctantly schlepped off to their own individual studios uncertain about what was expected of them and not knowing exactly what to do without an assignment. I sensed in them a kind of terror of the unknown. The point for me was clear—artists initiate, architects respond; John Hejduk and precious few others notwithstanding. Without a teacher, a client or at least a program and a studio critic, those architects at the academy were, at some level, unsure as to how to come to grips with their own design work.

*   *   *

One of the many perks of residency at the American Academy was to be included on educational journeys both in and outside Rome that were genuinely edifying visually as well as being informative. A day trip to Tivoli revealed characteristics about the Emperor Hadrian that would not have been immediately apparent had not scholars familiar with his life accompanied us on that trip.

I learned that the Canopus at Tivoli, a colonnaded-roofed structure, was oriented east-west with a single wall running the entire length so that the philosophers who advised Hadrian well could warm themselves in winter by strolling on the south side in the warmth of the sun and cool themselves in summer

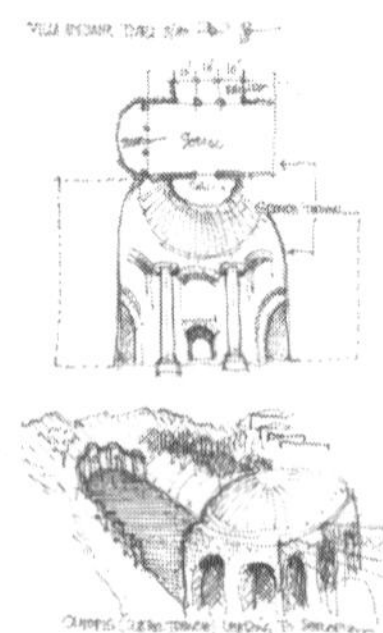

Ink-on-paper sketch
of Hadrian's Canopus,
Tivoli, Italy, 1980

Ink-on-paper sketch of the Pantheon, Rome, 1980

Ink-on-paper sketch of
Poikile and Triclinium,
Tivoli, Italy, 1980

by walking in the shade on the north side. Contradistinctively, the philosophers who advised Hadrian badly were destined to traverse the Canopus on the "wrong side" of the wall in those two seasons.

It was fascinating to compare the great interior space of the Pantheon in Rome with Hadrian's own bedroom in Tivoli, which incidentally had the identical diameter.

The Poikile and the Triclinium, both of which represented advances architecturally and technologically, were far ahead of their time and were fascinating formally as well as symbolically. The Roman's use of hot, tepid and cold chambers at the end of the day gave rise to methods of radiant heating and cooling used effectively in later millennia. Structurally, the inception of the arch and the vault would expand both architectural and structural vocabularies far into the future.

A particularly memorable trip was the James Ackerman-led week-long tour of the Veneto culminating in studying the Palladian cathedrals in Venice. The journey began in a somewhat terrifying way when I volunteered to drive the academy's VW bus through capricious Rome traffic. Once I realized that "mine was bigger than theirs," driving that behemoth became a piece of cake while minuscule European cars scattered like so many bugs before me.

It goes without saying that to be escorted through Palladian villas by the premier Palladian scholar of the time produced enormous benefits for all on that expedition. The villas Emo at Fanzolo and Barbaro at Maser by Palladio, as well as others by Sansovino, were well worth the trip all by themselves. But the highlight of that academy trip was being exposed to the two major cathedrals in Venice, both of which were also designed by Andrea Palladio. San Giorgio Maggiore and Il Redentore were magnificent.

It became eminently clear how much of a debt both British and American architects owed to Palladio. Thomas Jefferson's University of Virginia campus is but one example of Palladio's influence on subsequent generations of architects,

Ink-on-paper sketch of Palladio's Villa Emo, Fanzolo, Italy, 1980

Ink-on-paper sketch of Palladio's Villa Barbaro, Maser, Italy, 1980

Ink-on-paper sketch of Palladio's San Giorgio Maggiore, Venice, 1980

Ink-on-paper sketch of Palladio's Il Redentore, Venice, 1980

many of whom slavishly reiterated what Palladio initiated, and a small coterie of designers who extended the Palladian language brilliantly.

*   *   *

Three glorious months passed before I realized how brainwashed by the classical language of architecture I had become. To live in Rome for a considerable period of time in the presence of the level of scholarship that was available at the American Academy was to be exposed to an inescapable dose of classical enculturation with the authority of virtually two preceding millennia. I reluctantly returned to Chicago only to discover that Rome had damn near ruined whatever modernist sensibilities I had possessed before my halcyon hiatus in Italy. It took years of conscious renunciation on my part to re-embrace the modernism common to the place of my birth, Chicago, arguably the most modern city on earth.

Before my return to Chicago, however, there were some moments at the American Academy that transcended style, history or even culture. For example, when we traveled with Professor Jim Ackerman on a day trip to the Palazzo Farnese at Caprarola with a claque of his Harvard graduate students who were art history majors, the would-be historians clucked approvingly at Jim's encyclopedic knowledge of the oil paintings hung on the wall as they followed him up the great winding staircase of the palace. They all but missed the poignancy of the 6-foot 2-inch professor effortlessly carrying his disabled wife Mildred in his arms without any fuss or pretentiousness as he devotedly described the pictures along the way. Margaret and I, on the other hand, were especially touched as we watched the sympathetic way that he dealt with what those less sensitive might have perceived as an awkward situation. Jim made certain that Mildred received the same information that he imparted to his students. We admired Jim immensely for his humanity that was intrinsically tied to his social values through the vehicle of art history, and valued his friendship.

Staying at the American Academy meant "going down the hill" on sunny days to indulge in the incredible ambience that is Rome. Bramante's Tempietto at the Spanish Academy was just below McKim's neo-classical pile and we often stopped to study its proportions at close proximity

Jim Ackerman lecturing to art history majors, with his wife Mildred, at the Olympic Theatre, Vicenza, Italy, 1980

Ink-on-paper sketch of Michelangelo's Campidoglio, Rome, 1980

Ink-on-paper sketch of Bramante's Tempietto
at the Spanish Academy, Rome, 1980

even as we soaked in that special Roman sun as it enflamed the terra cotta of buildings around and about the Tempietto. At the bottom of the hill and across the Tiber we always strolled through Michelangelo's extraordinary civic space at the Campidoglio.

My 19-year-old son JJ visited us at the academy that year during his spring break from college. Just finishing his teendom and a sophomore at the University of Michigan, he was somewhat cowed by the less-than-self-effacing Rome prize scholars in residence. Attempting to keep him entertained, I suggested we attend a classical clarinet concert after dinner one night to which he reluctantly agreed, only to discover that the soloist, Bill Smith, was *Playboy* magazine's jazz clarinetist of 1980, and JJ, being the jazz buff he has always been, did a one-eighty realizing that the American Academy was in fact a hotbed of multivalence in ways that he could never have anticipated, but personally appreciated.

Later that summer, I was invited to participate in my second Venice Biennale. Entitled "La Presenca de Passato," the design and installation of my own pavilion, the *Facciata* of which was informed by the research I had done in the academy's library on the work of the 17th-century stage-set designers the Bibiena Brothers. After the festivities, a group of us including Margaret and my daughter Tracy, who had been with us at the Venice Biennale, boated along the Brenta Canal traveling from Venice toward the Palladio-designed La Malcontenta villa to have lunch with the descendants of the original owners. The Foscari's were then teaching architecture and urban design at the University of Venice.

The entrance to the author's
pavilion at the Venice
Biennale, 1980

One of the locks near the villa was under repair, forcing us to negotiate the remaining distance on foot. As we marched along the embankment, we all listened to a drawn-out squabble between Bob Stern and my then 15-year-old daughter Tracy, the substance of which has long since been forgotten.

I produced a sketch of the classical side of that brilliantly schizoid villa that faces the Brenta Canal. Its opposite rural façade faces the estate's farm land. Our hosts served a most delectable pranzo that we enjoyed immensely as we sat on that magnificent lawn next to the canal. Among the many sketches I made that glorious July while we languished in Venice was of Palladio's *Salute* on the other side of the Grand Canal. An iconic image that fascinated architects for centuries, I was simply one more who made that pilgrimage by vaporetto and recorded the results in my sketchbook.

*   *   *

Three other trips to Italy were instructional in ways that I couldn't have anticipated. In 1984, Margaret and I stayed once again at the American Academy, this time traveling westward from Rome with the academic Tom Schumacher to the Roman burial ruins at Cerveteri on the west coast of Italy. Like other architects before me who traveled to those remarkable 1st-century funerary monuments at the edge of the Mediterranean, we were predictably captivated by the hierarchies at work that represented the various economic and social strata in the cross-section of Roman life that was replicated after death. From the very rich who had their own villa-like monuments to the middle class who were buried in apartment-like tiers, to the poor common folk whose ashes were almost cavalierly buried in urns scattered around the floor of certain chambers, the entirety seemed in some ways to be fashioned after town planning principles that reflected the less-than-democratic social hierarchies of the Imperial Roman era.

In particular, I was entranced by polygonal versus orthogonal street layouts, to say nothing of columns centered on openings. It struck me that some 1st-century Romans harkened back to an era of pre-Hellenic architecture that displaced "space" with the mass of a centered column for those who had passed from this planet.

In my own case, because of my personal interest in "the architecture of death," the funerary monuments that I visited by the seaside of the Mediterranean meant a great deal to me. This hierarchical city of death produced an insight into what life both before as well as after death meant in Roman times.

*   *   *

The following year, yet another Roman excursion highlighted a tour of the remarkable Villa Pio in the

Vatican Gardens. Pirro Ligorio's brilliant tour de force is one of the most beautifully proportioned buildings that I have ever seen. Irrespective of the time that it was brought into being (1558), it is arguably one of the most modern structures in Rome. Meanwhile, since we were staying at the academy, John D'Arms persuaded Margaret to redecorate the Villa Aurelia gathering antiques from musty treasures stored in various attics throughout the property.

Finally, in 1990 I led a tour of the newly founded Midwest Associates of The American Academy in Rome first to Florence and then later to Rome, where I continued to execute a number of travel sketches. Visiting the Ponte Vecchio, like so many before me I was entranced by how the small-scaled shops that lined both sides of the bridge over the Arno River became a city in miniature hovering over the water. Another "must–see" in Florence is Brunelleschi's Pazzi Chapel. Ostensibly rudimentary, the chapel is a brilliant Renaissance study in understatement.

A highlight of the subsequent trip to Rome was Borromini's Piazza di S. Ignazio (1727). The architect's use of concave form to establish an embracing edge to this important civic space is still a radical urban design maneuver that at once welcomes one to the building, while adding the urban context by providing an inhabitable space that would not have otherwise existed but for the building's concavity.

Years later, on the occasion of delivering a lecture at the American Academy and when asked about my fascination with the dialectical affinity that I had suggested existed between Chicago and Rome, I observed that the relationship between the two cities was that, "While Chicago was the most modern city on earth, Rome was all the rest." I was well aware of Rome's authority since it was "all the rest" that had seduced me into the well-known trap that history plays upon us all by privileging itself, at least in terms of the longevity of antiquity that is lorded over one's own contemporary era.

More importantly, in my case, it suggested that historical forms are in some way more important than the modernist sensibilities of our own epoch as interpreted by the short time span that Chicago architects have had to deal with the subject. But as the decade of the 1980s came to a close, two projects half way around the world from each other came my way, each one playing an important role in extracting me from the trance of historicism.

The first was an exhibition I had been asked to curate on contemporary Chicago architecture and its antecedents. In addition, I was also asked to design the installation of that exhibition at the Gulbenkian Foundation in Lisbon, Portugal. Without my architectural residency in Rome, I doubt that I would have

Ink-on-paper sketch of Palladio's Villa Foscari (Malcontenta) on the Brenta Canal, 1980

Pranzo (lunch) at Malcontenta, 1980

Ink-on-paper sketch of Palladio's *Salute* on the Grand Canal, Venice, 1987

Margaret McCurry and the author in Rome, 1984

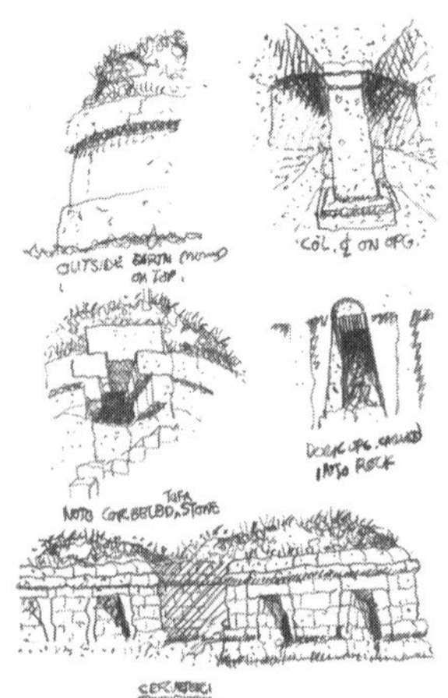

Ink-on-paper sketches of the cemetery at Cerveteri, 1984

Ink-on-paper sketch of the Villa Pio at the Vatican, Rome, 1985

Ink-on-paper sketch of Brunelleschi's Pazzi
Chapel, Florence, 1990

Ink-on-paper sketch of the Ponte Vecchio,
Florence, 1990

Ink-on-paper sketch of Borromini's Piazza di S. Ignazio, 1990

understood Chicago's multivalent architectural traditions in just the way that I did. Nonetheless, I was able to expand my palette for the exhibition installation far beyond the classical language of architecture that I had been so swayed by simply by being in residence at the American Academy. In any case, I had always understood that American architecture was a hybridization of European precedent and that Chicago, as a kind of capital of the Midwest, was the jewel in the crown of the understanding that melded preordained European forms with the American heartland.

The gallery given us for the Gulbenkian Foundation exhibition was linear, entering at the right end and exiting at the left. It didn't take more than the directionality of the space to encourage me to propose that the right-hand side of the space could be construed to be the conservative right-wing tradition in Chicago architecture that was exemplified by Howard Van Doren Shaw, David Adler, Daniel Burnham and so many others determined to extract themselves from the adventures of modernism. As one moved through the exhibition from right to left, everything became progressively more sparse and in due course "deconstructed," including the architecture of those more radical left-wing designers that were included as influenced by Louis Sullivan and Mies van der Rohe and architects of later generations who continued to hybridize or deconstruct the work of their modernist predecessors.

Organizing contemporary architects into stylistic categories based upon their formalistic preference didn't necessarily endear me to my colleagues in Chicago who perceived themselves individually if not indeed idiosyncratically original, but the exercise helped to clarify my understanding of the crossbreed qualities of the architecture of that part of the United States not immediately subject to foreign precedent.

No matter what stylistic camp one denies being a part of, the truth is that there are no originals in North America; only hybrids. Simplistically, originals tend to be found in countries composed of undiluted populations. From its inception, at least after European settlers displaced American Indians, the United States has been composed of a diverse and evolving population. This multi-ethnic population produced contextualized structures representing real, not imagined, formal variegation. That diversity is what makes American architecture inimitable and open to possibilities based on the latest émigrés that come to our shores bringing with them their formal antecedents, at least those most clear to them in their mind's eyes.

*   *   *

After the opening of the exhibition in Lisbon, Margaret and I leisurely motored northward to the Spanish border at Santiago di Compostela. Portuguese vernacular architecture is to my way of thinking more

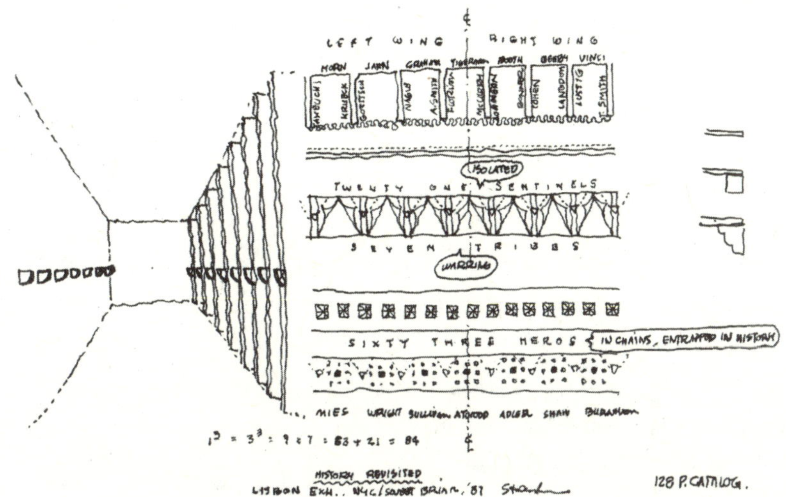

Ink-on-paper conceptual sketch of the Gulbenkian Foundation exhibition installation design, Lisbon, Portugal, 1987

Finished Gulbenkian installation (right wing), Lisbon, Portugal, 1987

Finished Gulbenkian installation (left wing), Lisbon, Portugal, 1987

reductive than its Spanish counterpart. Like Ireland is to England, Portugal is more rudimentary and to the point than its neighbor to the east. That doesn't mean that Portugal's architecture is of greater value than that of Spain's, just more basic.

Traveling through Óbidos, Alcobaça, Guimarães and Valença before arriving at Santiago, where we dined on the finest *angulas* (baby eels) in memory fresh from the Bay of Biscay with the smiles still on their faces, we witnessed architecture at its most direct and, at least simplistically speaking, devoid of decoration.

*   *   *

The second project on which I labored near the end of the 1980s was located in Fukuoka, on the south island of Kyushu in Japan. Not dissimilar from the IBA social housing project in Berlin that I had worked on a bit earlier although more experimental, the Japanese government filled in a discrete portion of the Sea of Japan and, through Arata Isozaki's influence, the Fukuoka Jisho Bank invited a number of architects from different countries to participate in the first stage of a multi-phase scheme, the design and construction of which they referred to as the Momochi Housing Project.

Unlike the IBA Berlin scheme of the mid-1980s, the Momochi project didn't particularly have a social dimension to its program, but nonetheless it was understood by its participants as being an opportunity to flex our muscles from a formal, as well as from an ideological point of view. It personally gave me an opportunity to preview theoretical ideas that I was working on even as I was writing about them.

While working on the Fukuoka project, I was concurrently writing *The Architecture of Exile*. I noticed that many congruencies between Western and Eastern cultures existed in their self-similar religious beliefs about origins. Buddhism and Kaballah are both spiritually mystical interpretations about how life began. Both beliefs seek what in Buddhism is referred to as nirvana, a state wherein one finds harmony between subject and object. While Kaballah is informed, at least in part by a particular mystical interpretation of numerology, Buddhism's mysticism relates more innately to the self bereft of the symbolism found in Kaballah.

Needless to say, given the precedent of Frank Lloyd Wright's personal interpretation of what it meant to work in Japan, I was fascinated by whatever connections I could make. That understanding inspired me to seek a unique expression for my project; it was simply not just another project there for me to work out my latest formal predispositions.

More to the point, I was at a stage where my formal ambitions intersected with my theological interests. Both were leading me to the design of grids and their relationship to scaffolding. I began to be interested

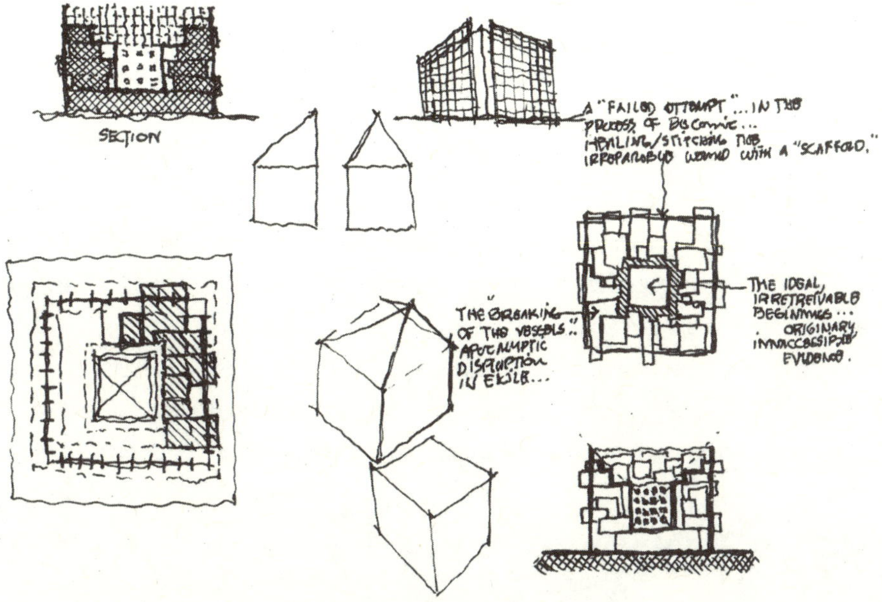

Ink-on-paper conceptual sketch of the Momochi Housing project, Fukuoka, Japan, 1989

Finished Momochi Housing project (Michael Graves' project on left), Fukuoka, Japan, 1990

in expressing the relationship between the provisional qualities required to build over and against the permanence conventionalized in the built realm. My interests in scaffolding also gave me an opportunity to erode the grid to avoid casting a shadow on the neighboring building to the north. By deploying a grid, I was also able to reflect on Japanese interests in modernism.

I was also stunned by the level of craftsmanship available to those of us working in Japan. Nisei workmen were even more precise than their vaunted German counterparts, and while the project was to be erected far from Chicago, the joint Asian and American technological sophistication reduced that distance palpably.

Atrium in the Momochi Housing project, Fukuoka, Japan, 1990

3

ARCHITECTURE AND THEOLOGY

By 1982, my son JJ, by then a recent graduate of the University of Michigan, was living with Margaret and me on the top floor of our Mies van der Rohe-designed Chicago high-rise. As he embarked upon his own life's unpredictable journey, JJ observed the trials that Margaret was facing in her five-year-old independent architectural practice. Margaret's 11-year apprenticeship at SOM may have helped to sharpen her already well-developed sense of connoisseurship, but being a part of a very large office didn't prepare her for the isolation of being a single practitioner particularly when that practice was operated out of an all-glass, west-facing, 425-square-foot efficiency apartment in which she lived as well as worked before we were married.

JJ suggested that Margaret join me as a partner in my own architecture firm, and given my penchant for instant decision-making in lieu of due diligence, I hastened to agree—never acts of omission, only acts of commission. This time my spur-of-the-moment decision was, in hindsight, incontestably the right one. In the same spirit that Margaret and I can be perceived as an odd couple in life, we are much the same in our architecture partnership. Rather than debate our very different design philosophies, when we joined each other in practice we were, and still are the masters of our own separate design destinies. By all rights our firm might just as well be called Tigerman "or" McCurry, since that's the way that we approach practice, which is to say that the two of us operate on different architectural wavelengths. While we crit each other on a regular basis, each of our projects in general evolve more or less autonomously.

During one particular Christmas holiday season, an attempted break-in at our office resulted in no greater damage to our studio than scratching the glass entrance door to our offices, resulting in the need for replacement glazing. We had to have a new sign painted on the door, and without consulting Margaret, I modified the language of the firm from Tigerman McCurry Architects to Tigerman or McCurry. Margaret didn't notice my prank for some months, but when she did, she was not entirely enchanted with the change and just like that, the "or" was eradicated leaving the space in which the word had formerly existed as a kind of encoded palimpsest of our contradictory practices.

Our differentiated careers are the architectural equivalent of a hedge fund; if I'm not busy, she is and vice versa. Nonetheless, our apparent independence one from the other is tempered by proximity. I would be disingenuous if I didn't admit that the juxtaposition between us didn't have an impact on each of our individual architectural production. Margaret is more sophisticated than I am, while I am somewhat more impulsive than she is. Margaret's studiousness has to be seen in the context

St. Benedict's Abbey Church, Benet Lake, Wisconsin, 1971

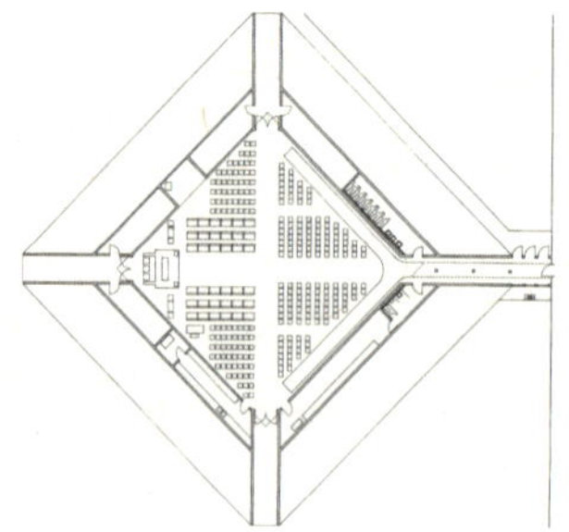

Ink-on-paper drawing of
St. Benedict's Abbey Church
floor plan, Benet Lake,
Wisconsin, 1971

of my being a "quick study." There is much to be said about Margaret's thoroughness, yet it is her work ethic that at the end of the day separates her from many other architects. She doesn't engage in "marketing," and is utterly unaffected by "branding," she simply puts her head down and works. She even cheats on her timecard downward because she feels it is inequitable to clients that she is slower than she feels that she should be. I can't begin to say how much I admire her work ethic and the resulting focus that forms the basis of her well-thought-through architectural production.

*   *   *

During the difficult decade of the 1970s, I was befriended by a Benedictine Monk, one Father Michael Komachek, an architecture buff who consulted from time to time with the Chicago Catholic Archdiocese on matters of design. Father Mike was a life-long devotee of the design work of my old friend, the Chicago architect Ed Dart. One of Father Mike's duties at the archdiocese was to recommend architects periodically for projects administered by the Catholic Church in the greater Chicago region. The priest became interested in my work and put me on a list together with Ed Dart, SOM's Walter Netsch and Harry Weese to be interviewed to become the architect for the St. Procopius Benedictine Monastery in Lisle, Illinois, where he was in residence. The commission went to Ed Dart, but Father Mike's and my emergent friendship led to an interview with another Benedictine monastic community northwest of Chicago at Benet Lake, Wisconsin. St. Benedict's was considering developing a master plan and building a monastery church on their large picturesque property, which straddled both sides of the Wisconsin and Illinois border.

A far-reaching program for St. Benedict's Abbey had begun some years earlier when its first abbot administered the design and erection of their overly ambitious original building that had a capacity for housing 144 monks. A three-story stone edifice with double-loaded corridors—an over-scaled rectangular parallelepiped with a straightforward gabled roof—sat in a peaceful prairie on a plateau high above Benet Lake. When I arrived, there were just 12 monks residing in their exaggeratedly outsized building, but with no monastic church occupying the grounds.

During the course of my interview with them, I was informed by the monastic architectural selection committee that one of the architects being considered for the commission was the son of the original architect who had designed the less-than-inspiring existing monastery. I suggested to all assembled that since it was their religion that espoused the concept that "the sins of the father are borne by the son," it seemed only sensible that the descendent architect should, by all rights, be given the commission. At once disarmed and amused, the monastic community awarded me the project on the spot.

As I began analyzing what might be a logical master-planning approach to their scenic campus, I often chided them that since they had hired a Jewish architect, Jesus might go "underground." Whether that flippant comment influenced my eventual design of the partially buried chapel with its crown-of-thorns roof structure that defined the central sacred space of the chapel is a matter of conjecture.

It soon became evident that the exceptionally large original monastery might benefit from a design that would reduce the presence of that facility to a more modest incursion on the beautiful plateau overlooking Benet Lake.

As I began to spend weekends living at the abbey and working with members of the monastic community to develop the program for master planning their property, I observed that the individual and collective characteristics of the faithful were unlike those of people that I knew "on the outside." Imperfect as some of the monk's lives might have been, I feel safe in saying that they seemed to be less flawed than most of the people I had known in the secular world. Perhaps my perception of the calmness of the place was because the Benedictines were categorically men of faith, but I don't believe that that's the entire story. I marveled at the tranquility that pervaded the abbey sitting on that serene plateau high above the glacial residue known as Benet Lake, and the peaceful aspect of the few souls who resided there whose lives were focused on practicing their faith during their many daily worship services.

I was going through a divorce from my first wife at the time and being in regular attendance at the monastery had a considerable impact on the way in which I conceived the design of the monastery chapel. The early 20th-century German architect Rudolf Schwarz once stipulated in his book *The Church Incarnate* that "the church is the people." In the spirit of that seminal book, introduced by Mies van der

Rohe, I configured the aisles of the sacred space in the form of a cross, so that as the lay congregation queued up for communion, the crucifix-like configuration of the aisles symbolically as well as literally prepared them for their destination.

The pinwheel-orientated roof structure became a kind of crown of thorns sitting upon a significant earth berm that otherwise defined the building thereby reducing the scale of the chapel as well as that of the original structure since they were both juxtaposed perpendicularly to each other.

Like their Cistercian predecessors that Umberto Eco describes so tellingly in his book *The Name of the Rose*, I felt that the Benedictines were just trying to get "it" right, whatever "it" was. As I sat during services listening to their repetitious chanting which punctuated their prayers, I began to gain an insight into the concept of value-through-habituation, and what that idea might mean as an ethical elaboration of architecture. My epiphany was something of a revelation, but it also reified the Socratic concept of habituation. This linkage between theology and philosophy in terms of value was, for me, of great consequence and was to be influential in the way that I approached architecture from that point forward.

Over the course of the decade that I spent laboring over programs given me by the monks of St. Benedict's, in addition to the chapel, I also designed a printing plant for the monks and did some remodeling work on the original building. My many years of working with them had great significance to me in ways that I was not able to determine at the time; it took me decades to thoroughly comprehend what had transpired in my association with the monks of St. Benedict's Abbey.

*   *   *

From a religious perspective, my grandfather's influence reemerged into the changing equation of my current life. From that point forward, I became increasingly cognizant that my writing, my teaching and, ever so slowly, my architectural practice began taking value-through-habituation into account.

My early forays into habituation through repetitious tasks seemed to find a home in the larger context of my architectural production. One can potentially find serenity in reiteration, particularly if repetitious tasks are not done simply by rote, but with a modicum of thought about the nature of habituation and the value that it brings to one's life.

None of this is to say that my earlier, one-off, sometimes idiosyncratic work was bereft of value, and I stand responsible for all that came before. However, if there is one thing that was revealed to me, it is that the only constant in life is change-qua-change. Habituation is something one can turn to when overwhelmed continuously by unremitting change. In a very real sense, the concept of habituation has helped me to redefine my life as an architect in ways that I could not have anticipated.

*   *   *

My own personal problems notwithstanding, the decade bracketing 1970 was a difficult watershed for the United States as well. The Vietnam War, student revolutions and the assassinations of both Robert Kennedy and Martin Luther King all conspired to challenge our national sensibilities. In addition to the personal anguish that such events triggered, these incidents were also a shock to the commonweal. Individually and collectively, many of us were forced to come to grips with the unfamiliar, but at the end of the day crucial—if cruel—concept of loss.

For me, a paradigm of loss transmuted into hope and actualized in architectural terms is exemplified by the architect Maya Lin's Vietnam Veterans Memorial in Washington, D.C. dedicated to those Americans who died in that controversial conflict. Before that time, Americans had memorialized, if not precisely glorified war by erecting pictorially literal, columnar structures symbolizing victory in their verticality and generally light colored monuments to celebrate winning. What is so moving about the Vietnam monument is that it is a horizontal black wedge. It commemorates loss by listing all the war's victims chronologically as they fell on the battlefield on granite half buried in the ground with its arms pointing powerlessly, yet hopefully towards the Lincoln and Jefferson memorials.

However, one's faith might be tested, or contradistinctively however one interprets life in lieu of faith, one can be assured that change challenges both. My own familiarization with the concept of loss that emanated from failure repeatedly bombarded my own psyche; flunking out of school, divorce(s), periodically being fired by clients and ultimately being removed from a leadership position in teaching all conspired to personalize disappointment so that at the end of the day I was forced to internalize its meaning viscerally.

*   *   *

St. Benedict's Abbey was not my only opportunity to reflect about the conceptualization of forms in theological terms. I received a commission to prepare a master plan for St. John's, the Catholic chapel at the University of Illinois at Urbana-Champaign (UIUC). The chapel represented one wing of a U-shaped courtyard building, the other two wings of which were devoted to Newman Hall, the Catholic men's dormitory at UIUC. The brief was to fill in the open courtyard of the building thereby providing a much-enlarged narthex, which would simultaneously simulate an outdoor space that might be used for events both theologically and secularly driven. While the project was never brought to fruition, the joy that I experienced in being connected with the challenges inherent in a religiously based project in such an important arena, that arena being religion, wasn't lost on me.

Another significant theologically informed project that came my way in the 1970s was the National Archives Center for the Baha'i faith in the United States, although the interview process for selecting an architect did seem a bit odd. Four architects were invited and interviewed, the other three being persons of the Baha'i Faith. When I was informed of my selection, I couldn't quite understand the reasoning behind their choice until I attended the first programming meeting. This session occurred in the lower level of their existing, nine-sided temple in Wilmette facing Lake Michigan on the North Shore of Chicago. While waiting in the ante-chamber to join the meeting, I nonchalantly looked over an assortment of public relations material about the evolution of the Baha'i Temple. Various brochures included biographical information about its Canadian-born architect, Louis Bourgeois, whose year of death amazingly coincided with my own year of birth.

Remembering that my research on the Baha'i Faith had uncovered the fact that the death date of the founder of the faith, Bahá'u'lláh, coincided with his grandson's birth who then succeeded his grandfather as the head of the faith was a twist of fate that I couldn't overlook. The Baha'i selection committee never admitted that this strange coincidence informed their choice of an architect, but I could never otherwise explain their selection.

*   *   *

One of the first assignments from my new client was to send me to Haifa, Israel in April of 1977 to study the Baha'i facilities there as well as to get a better understanding of the origins of the Baha'i Faith. While in the Middle East, I extended the trip a few days by traveling throughout Israel to observe the confluence of so many of the world's religions. The journey culminated in Jerusalem significantly on Good Friday. Standing on the stones of the Via Della Rosa, I heard a mullah calling the Muslim faithful to prayer, even as I observed a devout Christian group retracing the steps Jesus took on his way to his crucifixion. Friday also being the Jewish Sabbath eve, the same Via Della Rosa was being crossed by Hasidic rabbis on their way to prayer at the west wall of the Temple Mount. It was the closest thing one could imagine to a stage set organized by Hollywood's Central Casting.

*   *   *

Informed by my earlier conceptual project for the premier exhibition of the Chicago Seven, my design for the Baha'i project was an attempt at cleaving an otherwise symmetrical, holistically derived form in order to express the difference between that which is finite and that which is eternal. Conceptually, one half of the cleaved form housed the program for the National Archives Center that was intended to represent finitude while the other half was composed of self-similar topiary-deformed trees representing the perpetual.

Rendering of Baha'i Archives Building, Wilmette, Illinois, 1982 (rendering by Merike Phillips)

The axis that resulted from the cleaving of the two halves pointed to the past represented by one of the nine sides of the original Baha'i Temple situated across Sheridan Road, while the axis in the opposite direction was oriented to the future represented by the sun rising in the east over Lake Michigan.

The project was never realized because of a legal conflict that arose between the Baha'i Faith and the State of Illinois as to the ownership of the property's riparian rights on the lake, but my work for the Baha'i Archives Center represented another example of cleaving so as to get to the heart of a project's core.

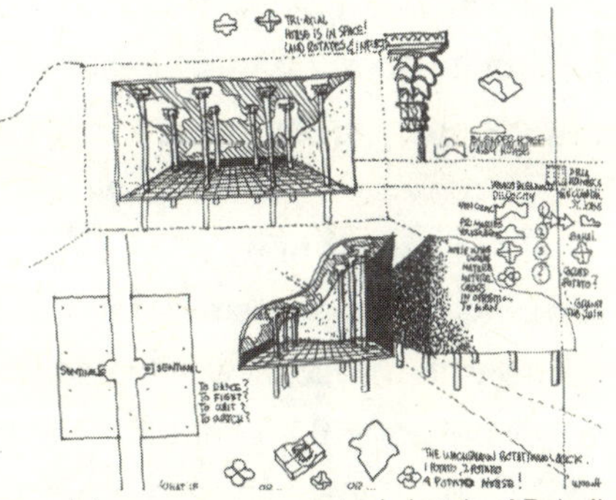

Ink-on-paper conceptual sketch of Baha'i Archives Building, Wilmette, Illinois, 1982

*   *   *

The reasons underlying my interest in cleaving otherwise totalized forms represented my unfathomable uneasiness about symmetry. Throughout my entire architectural career, my love-hate obsession with symmetrically derived form refers to the almost magnetic draw that symmetry has on many architect's approach to design of which my own susceptibility to such things I readily admit. Cleaving symmetrical forms was my simplistic way of acting out against history's authoritative hold on architecture which I accomplished initially by using symmetry to define buildings by axiality; I instead rent that same symmetry asunder.

I also think that my repugnance to unchallenged symmetry lies in my reluctance to be a part of that greater continuum which is, after all, the history of architecture itself. The challenge to symmetry is the work of an outsider insisting that he remain outside.

Even under Paul Rudolph's Yale tutelage that espoused modernism to the exclusion of all else, and much to my professor's chagrin I seemed to be incapable of conceiving asymmetrical or modernist forms on any sustained basis. I couldn't give the impression of separating myself from the authority of the human body with its biomorphic symmetry arranged about the vertical axis, and yet I was committed to work in the context of my own epoch with its modernist values.

Marcus du Sautoy's 2008 book *Symmetry* convincingly explores the reasons underpinning humankind's affinity for things symmetrical. Starting with how symmetry impacts the concept of "the survival of the fittest," du Sautoy threads his way through the natural forces that have driven symmetrical choices throughout the natural world. Bees seek out the most symmetrical flowers because they produce the sweetest nectar, while flowers seek the greatest symmetry so they can be pollinated effectively. In architecture, overarching symmetrical forms demand local symmetries to attain a kind of formal fulfillment, an architectural task few designers have accomplished. Suffice it to say that the biomorphic human physiognomy has an influence of some magnitude on those who would design.

To represent the era in which I was working, I felt compelled to rupture what I construed to be idealized form as otherwise expressed in balanced terms. This cleaving act represented my own perverse approach to modernism by rending a symmetrical whole rather than resorting to asymmetry as the more formally conventional way of rejecting classicism.

In the early 20th century, Frank Lloyd Wright was the first among many architects to "break out of the box" by escaping out from the corner, thus establishing modernism as a historically canonic, albeit disjunctive, perhaps even dysfunctional movement. Before the 20th century, the history of Western architecture is rife with examples of bilaterally symmetrical structures. It took the apocalyptic work of modernist architects working at the dawn of the 20th century to rupture that tradition once and for all. It can be argued that Frank Lloyd Wright was the progenitor of that activity. Before that time, the corners of buildings were increasingly reinforced visually so as to hold fast to axial antecedents. When Wright exploded the contained volumes of predecessor structures into bypassing planes, everything changed.

The disruption to which I am alluding is that virtually all historic movements—at least those that originated in Europe, with the exception of the Gothic one—were built upon the foundations of what preceded them.

The modern movement rejected historical antecedents in favor of the tabula rasa but, in contradistinctive terms, utilized predecessor forms as a kind of thesis to overturn. At the beginning of the 20th century, the idea of the clean slate was in the air in the arts specifically and in culture generally. Was it a coincidence that modernism coincided with the rise of the Freudian and Jungian psychoanalytic movements that at least identified, indeed tried to explain, if not to validate the conflictual aspects of the times?

Like the apocalyptic view of creation represented by the second part of Lurianic Kabbalah, Sheviret HaKelim, my conceptual foray into the disjunctive aspects of architecture occurred in part as a result of my working in an autodidactic sense on theologically based projects. Since I've never believed in coincidence, I began to focus on those aspects that I was able to coax out of my understanding of religion as it relates to informing architecture. This effort represented my emergence into my own personally defined arena of self-induced conflict.

To the extent that I thought about "conflict resolution," it reminded me of the tripartite Hegelian concept of first "thesis" which then employs "anti-thesis" as a straw man expected only to prove the thesis that was proposed in the first place, which was then ennobled by assigning the term "synthesis" to the process as if the original thesis needed a new name for legitimization. Simplistically, I rejected tripartite Hegelian concepts as representing a level of irrevocability—that is, closure that was contradistinctive to my (Jewish) compunction about interpretation as opposed to the more normative understanding that is connected with Hellenic-Christian precedent steeped in faith.

*   *   *

An entirely different and utterly unpredictable circumstance occurred as a result of another commission within the same typology. The Chicago-based reformed Jewish congregation Or Shalom owned property in a North Shore suburb. The congregation asked me to design their new temple in that location. Of course, I was aware that the site was located in a suburban neighborhood where there were literally no Jewish congregants (nor Jewish residents for that matter) and understanding that Judaism was a distinct minority in the United States in any case, I designed something of an underground structure that, together with an earth berm that was to surround the building, would virtually hide the structure from view. I thereby hoped to avoid making a statement that might be construed as contextually challenging, or in any way disruptive to existential precedent.

In spite of all my efforts to palpably reduce the presence of Judaism so as to not offend the burghers

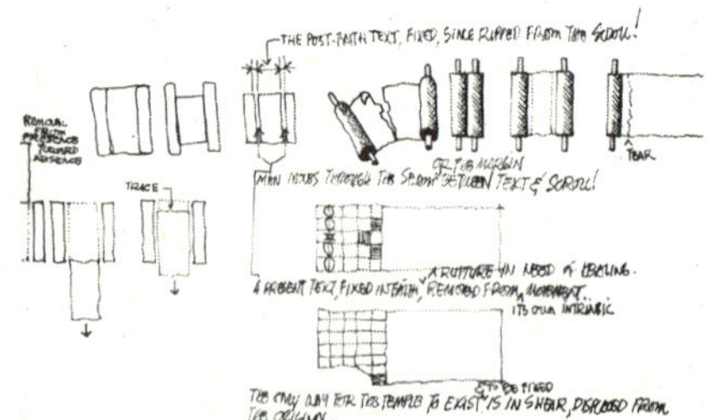

Ink-on-paper conceptual sketch of Or Shalom Temple 1, Metawa, Illinois, 1986

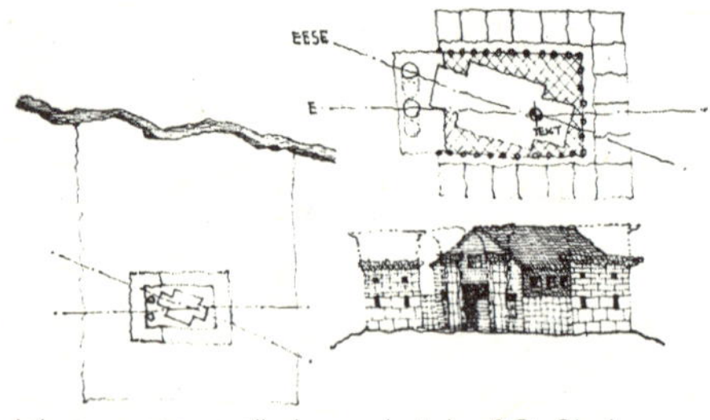

Ink-on-paper preliminary sketch of Or Shalom Temple 2, Lake County, Illinois, 1986

of that quiet community, the neighbors in that North Shore suburb nonetheless initiated a class-action lawsuit against Or Shalom to bar the congregation from proceeding with the project. I hoped that my client might fight the lawsuit because it was, to my way of thinking clearly defensible, but being a minority and not wishing to cause trouble, the congregation sold the property. Ever optimistic, they acquired another piece of property close by in an unincorporated section of Lake County.

There was an existing log lodge located on the new site which, when analyzed, was found to be oriented slightly east-southeast, or toward Jerusalem. It seemed appropriate to expand upon that building by linking it with a building addition oriented due east "in anticipation of a Messianic age," utilizing as a precedent the orientation common to the several Temples of the Jews prior to the coming of Jesus and afterward the several Diaspora that would alter the Jewish orientation to one that would face the West Wall of the Temple Mount.

At that point, the reformed congregation's young rabbi informed me that, "anticipating a messianic return was a bankrupt concept!" I never had a chance to challenge his argument since before I could get much farther with the design than a compelling study model, I discovered that our newest residential neighbors had filed yet another class-action lawsuit, and once again the congregation sold their recently acquired property to avoid a confrontation. It appeared to me that Nimby (not in my backyard) was alive and well in Chicago's North Shore suburbs.

In an act of frustration, the congregation finally agreed to a time-sharing arrangement in a building that housed a financially troubled Christian Science church located in an adjoining village. As a result, I was asked to design a movable bimah that could be put away during Christian Science services. At this juncture, I decided that it was time for me to retire from the fray. However, in an unusual twist of fate my two earlier designs were sought after by my friend Heinrich Klotz, the director of the Deutsches Architekturmuseum in Frankfurt am Main, Germany.

I had met Heinrich Klotz earlier when I donated 60 of my "architoon" drawings to his museum that were subsequently made the subject of an exhibition that ran from February 5 to April 24, 1988. Heinrich's museum was hosting an exhibition and a book on synagogues on the occasion of the 50th anniversary of Kristallnacht. The two Or Shalom unbuilt temples were the only Jewish projects that were not located in Germany that were cited in the exhibition and illustrated in the catalog.

In the context of Or Shalom, the conflict to which I am alluding was not only implied in its architecture, but in the entirety of a project designed for a religious minority. The rejection of the planned synagogue by neighbors in two separate locations denoted something onerous still at work in a society that makes a virtue of espousing the benefits of existing in a multivalent society that trumpets itself as a "melting pot."

*　　　*　　　*

Beginning in the mid-1980s, my writing as well as my approach to design was increasingly informed by an urge to link architecture with Jewish theology. My earlier discoveries about biblical measurement and its disjunction with Western measurement and the differentiated orientations that bracketed the coming of Jesus were sufficient to ensure continuing focus on Jewish theological precedent. The major outcome of these meanderings led to my writing *The Architecture of Exile*, a book that was published in 1988 and that attempted to explore the nature of those intersections. The work signaled my growing interest in connecting Old Testament traditions with an exilic American architecture. I dedicated the book to my grandfather whose death a half-century earlier virtually coincided with Kristallnacht,

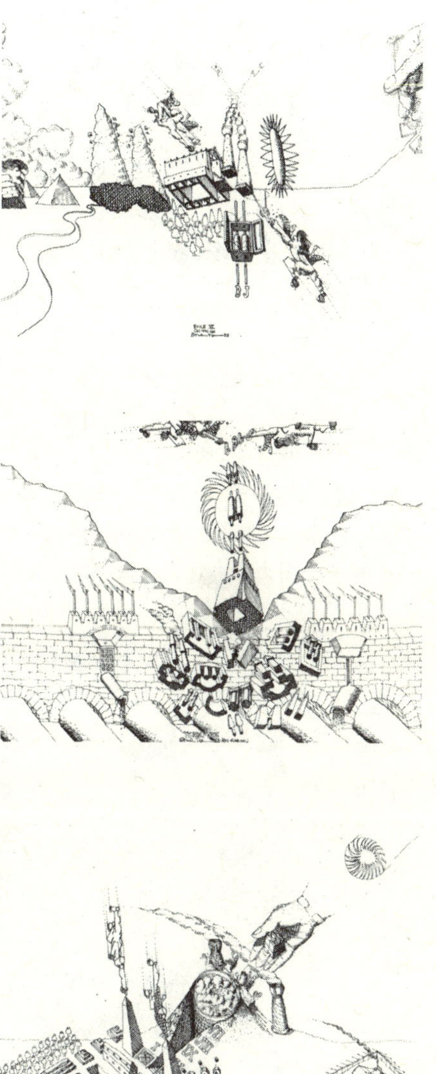

Ink-on-paper conceptual sketches for
The Architecture of Exile, 1983

205

From left: Tracy Tigerman, the author and Margaret McCurry posing for the cover of *Dorothy in Dreamland*, 1991

the notorious night in the late 1930s when Jewish-owned shops and synagogues in Germany and Austria were trashed. I also made a number of surreal-like sketches that I included in the book that graphically represented ideas that I was ruminating about at the time.

Drawings that were self-similar to those that I used in illustrating The *Architecture of Exile* started appearing as a result of my apparently random sketching that began in the mid-1970s and continue up to the date of this writing. These drawings are at once theologically inclined from a symbolic point of view and surrealistically informed from an aesthetic point of view.

A decade later, Rizzoli commissioned John Hejduk, Charles Moore, Bob Stern and me to write and illustrate four different children's books. Hejduk did a takeoff on Aesop's Fables, Moore did an iteration of *Beauty and the Beast*, Stern produced *The House that Bob Built*, and Margaret and my daughter Tracy collaborated on the text and I did the illustrations for *Dorothy in Dreamland*, a conflation of a number of children's fables that I wanted to revisit. All four were published in 1991 as the Rizzoli Classic Series.

*   *   *

At the same time that I was writing *The Architecture of Exile*, I was designing a small urban villa

Ink-on-paper preliminary drawing for the IBA project, Berlin, Germany, 1987

in Berlin as a part of the Internationale Bauausstellung (IBA) project. The program called for a three-story apartment building based on ideas about social housing with two flats per floor to accommodate working-class families. Trying to be both contextually and historically responsive, I modeled my design on Mies van der Rohe's first Perls House in Berlin of 1909 and stained the stucco of the apartment building in the saturated red, yellow and black colors of the German flag. In the spirit of cleaving an otherwise holistic form, I then proceeded to split the building into two distinct parts, rationalizing the severed center space by making it into a "winter garden" (such a function was common to

German housing). I would be dissimulative if I didn't admit that the principal agenda for my interest in cleaving the project was my unwillingness to produce a holistically inclined form in a country that had a less than laudable history by its brutal depletion of its Jewish population.

The architects who also participated in the Josef Paul Kleihues-organized IBA represented an international assortment including: Raimund Abraham, Peter Eisenman, John Hejduk, Charles Moore, Robert A.M. Stern and me from the United States; Zaha Hadid and James Stirling from the United Kingdom; Hans Hollein from Austria; Rob Krier and O.M. Ungers from Germany; and Aldo Rossi from Italy.

Entrance to the Winter Garden of the IBA project, Berlin, Germany, 1988

*　　　*　　　*

Beginning within a decade after the end of the First World War and every 30 years thereafter the German government sponsored the building of experimental social housing. The first development, the so-called Weissenhofsiedlung was organized in 1926–27 under the leadership of Ludwig Mies van der Rohe who master planned the site and included a number of mostly young, progressive European architects. The selected German architects included Mies himself as well as Peter Behrens, Walter Gropius, Hans Scharoun, Hans Poelzig, and Bruno and Max Taut. Le Corbusier represented France and J.J.P. Oud, the Netherlands. Sited in Stuttgart, Germany each architect was asked to design an apartment complex that would relate to each other contextually on a single sloping site.

Thirty years later in 1956–57, the German government sponsored the subsequent Interbau social-housing project, this time with a series of scattered sites located in Berlin. Each site was designed by a different architect. Again, a number of the most important European architects of the day were retained to create unique possibilities for workers' housing and included Alvar Aalto from Finland, Jacob Bakema from the Netherlands, Egon Eiermann from Germany, Le Corbusier (again), Oscar Niemeyer from Brazil, and Pierre Vago and Shadrach Woods from France. Thus, the most current IBA project had two well-known urban antecedents to draw upon.

*　　　*　　　*

On one particular trip to Berlin, I traveled into East Berlin through Checkpoint Charlie to visit Karl

Ink-on-paper sketch of Schinkel's Schauspielhaus

Friedrich Schinkel's Schauspielhaus. For me personally, this concert hall is one of the most sophisticated buildings built anytime, anywhere.

On that same trip, two Berliner architects and I traveled by car through East Germany to visit Schinkel's Sanssouci project near Potsdam. On the spur of the moment, we belatedly decided to picnic in a rural area not authorized by our visitor's permits. When the East German police discovered us, I was not as user-friendly as I might have been under more pleasant circumstances, which resulted in the three of us being arrested and paying a short visit to the local East German hoosegow.

*   *   *

The private-sector real-estate development community that sponsored the IBA projects held an annual spring celebration that took the form of a cocktail reception in West Berlin on the occasion of Ascension Day. Given that my first trip to Germany coincided with Passover, being a Jew in Berlin to some degree informed my conduct which resulted in behaving in not the most admirable way. I acted out that angst by showing up at one particular Ascension Day party inebriated, undiplomatically accosting every Berlin developer older than I was and aggressively wondering aloud where they all were 50 years earlier at the time of Kristallnacht.

In 1986, I gave a number of lectures in Austria and Germany. They began innocently enough in Vienna where Hans Hollein escorted us around town to see the work of Josef Hoffman. I had known Hans from his days as a graduate student at IIT where he studied with Mies and we had renewed our friendship at the 1976 Venice Biennale.

After my experience at the IBA party in Berlin, and immediately following our visit to Dachau, I began my Austrian and German lectures confrontationally by stating belligerently: "Ich bin ein Juden." Nonetheless, the talk I gave at the architecture school in Graz, Austria was very well received. The students invited me to a "new wine" tasting picnic near the Hungarian border the following day. Accompanied by the eccentric

Austrian architect Gunther Domenig, the picnic began festively enough but after a bit too much new wine and blood sausage, one of the professors from the architecture school in Graz, who had lost an arm while serving in the German Army in the Second World War, began behaving obstreperously. In hindsight, I think that he was offended by my rock-star-like reception from the students. At one point he confronted me by saying that he was certain that he knew "who was my hero." When I asked him who that might be, he said "Abraham." Stung by the comment and not to be outdone, I responded by saying that "I knew who his hero was," adding "Adolph Hitler!" The picnic erupted in fierce argumentation between students and faculty, which is to say between generations.

On that same trip, Margaret and I came across a six-and-a-half-foot tall, mystical, John Hejduk-like architect in Salzburg who both lived and had his studio at the time in a converted convent. Hans Hiegel told us the story that centuries earlier, one of the young novitiates had become pregnant and the nuns buried her alive within the walls of the convent, and that he heard her wailing from time to time in the night. Given that my grandparents had come from the Borgo Pass in Transylvania, I listened to his tale in awestruck silence. It seems as if I would never have a Middle-European experience that even approached "normalcy."

*   *   *

Once I began working on a topic as rich in interpretive possibilities as is the Torah, I found it impossible to even think of terminating my exploration of linkages between architecture and the Jewish religion. Thus, after publishing *The Architecture of Exile* I began a book (still in progress) entitled *Failed Attempts at Healing an Irreparable Wound*. The theme of the book was related to a relationship that I felt needed to be explored between Jewish mysticism and architecture. I was still searching for the unheimlich in architecture and whatever mystical experience that came my way was grist for my explorations. I believed that a strong connection existed in need of amplification between the three portions of Lurianic Kabbalah and an ethical approach to architecture with the exception that my interpretation of the final element of Kabbalah—Tikkun—was unresolved as opposed to the more conventional understanding of Tikkun, which is a bit too much like Hegel's "synthesis" that neatly wraps up the ostensible conflict between "thesis" and "antithesis."

While it's obvious that being Jewish was a rationale to motivate me to pursue this connection, I'm equally certain that my concern about ethics-qua-ethics was also a factor in this pursuit, since over time I was writing, editing, presenting papers at conferences and publishing several works on the subject. Later, as director of Archeworks, I initiated a number of publications to present the work of independent scholars who were commissioned to address the subject of ethics generally, and architectural ethics in particular (*The Archeworks Papers*; vol. I, nos. 1–5).

Each publication also featured three responses, or challenges, to the presenter's interpretation of ethics. The entire set was modeled after the *Norton Papers* that were first delivered as lectures on a particular topic at Harvard College, occasioning subsequent publication. Since it was in my interest that Archeworks position itself as a forum so as to debate one major topic in depth, the ethics course that I taught at the school set the tone for the lecture presentations as well as for the resultant publications that I then edited.

*   *   *

I entered a conceptual competition in 2003 whose brief involved the designing of an "interfaith" sacred space. Situated in an imaginary urban square in San Francisco and organized around an enclosed space with an inaccessible center, the circular design of my particular entry featured 12 differentiated religious structures on the perimeter of the site that expressed 12 collectivized congregations each orientated correctly within the traditions of each particular religion. Moving in toward the center of the space were 12 smaller self-similar structures, each with six private chambers for individual meditation in the spirit of Søren Kierkegaard's concept of isolated communion with one's own individual deity. The unattainable center with its implied axis mundi suggested an equal inaccessibility to a universal divine being and was in any case, reflective of my fascination with what John Hejduk always referred to as "aura," or "ineffability" in architecture particularly in the context of cloistered or contained space where a sacred center is denied to humankind.

My attitudes about the subject of an interfaith ministry were informed in some measure by my wife Margaret's cousin Elizabeth Stout who was herself an interfaith minister. Not entirely unlike the Baha'i Faith's ex Shia Islamic antecedents, interfaith theological issues are ecumenical in intentionality.

*   *   *

For almost 20 years, I have on occasion attended by invitation faith-based conferences where I have presented papers focusing on how Judaism innately infers temporality as a viable architectural option in contradistinction to the summa of Christian theology—time without end. Informed by Jewish interpretation versus Christian faith my continuing interest in unresolved dialectics continued apace.

Testing such ideas by alternating between writing and designing in the context of theology as it relates to architecture has been useful in bringing to consciousness, if not always clarifying, my views on these matters. It was fortuitous that my teaching career closely paralleled the development of my practice, otherwise I might never have been exposed to what I now consider as crucial to both, that is the interaction between them.

I rationalized my dualistic position by engaging in both give-and-take aspects of the "giving" of teaching and the "taking" of practice. I always thought that teaching was defined by the generosity of time and effort that it took to empower students to perform effectively in the field, whereas practice was defined by satisfying the self intellectually and ideologically through a formalistic set of conditions that one might have been engaged in at the time.

*   *   *

With respect to the teaching side of the equation, I wrote *The Architecture of Exile* during the period that I was a visiting critic at Iowa State. Students in the design studio in which I taught produced ingeniously conceived and impeccably drawn interpretations of the biblical text that was described in the theological language of the Torah, describing the constituent architectural features of the Temple of Solomon by continuously referring to measurement which is rife in the Temple's description in the Bible. The students scrutinized the biblical text on the subject (1 Kings: 5–9 of the Old Testament) as well as making visual interpretations of the Prophet Ezekiel's "temple in an anticipation of a messianic age."

Ames, Iowa had precious little in the way of distractions, thus I was free to write without interruption. The ISU architecture dean at the time, Ken Carpenter, was extraordinarily supportive with my meanderings, the Iowa State Library in the middle of the "Bible Belt" had a superb section on religion, and frankly, I spent that wonderful semester in peace, content to develop my text at will which, upon completion, I then dedicated to my grandfather.

Writing *The Architecture of Exile* was self-revelatory since it covered ground that I felt had not been substantively gone over by others. Penning the text also represented a purging of feelings that I had about Judaism in the context of architecture that I felt were important, in part I'm sure because of my grandfather's Torah studies.

*   *   *

The year 1992 represented the 500th anniversary of Columbus's expedition to the Americas, as well as the 500th anniversary of the Spanish Inquisition where Moors and Jews were given three notorious choices: convert to Christianity, be subject to expulsion from Spain, or die.

For that seminal year in Iberia, I was one of five architects asked to participate in an exhibition of planning and architecture in Madrid that was organized by then expat Martha Thorne, who represented Madrid's Ministry of Culture. She had been asked to curate a number of urban design interventions, each one of which was to be situated on a metropolitan site of one's own choosing from a list prepared by the Ministry of Culture and each of them overseen by Martha.

I traveled to Madrid to visit the proposed locations, but was dissatisfied with any of the several sites that the Ministry of Culture offered us. In lieu of any of the pre-selected sites, I picked the location of the Almudena, Madrid's most significant cathedral. When I analyzed the site I was flabbergasted that such a seminal place of Christian worship seemed to have been orientated erroneously. The cathedral's entry faces north rather than west, apparently paying homage to the Royal Palace with its statue of King Philip guarding the entrance rather than acknowledging divinely inspired Cistercian monastic precedent with its canonic east/west orientation.

Since the 10th century, which saw the rise of Cistercian monasticism, virtually all European cathedrals that looked to monastic antecedents for inspiration have been oriented to the west. The exception to the rule is St. Peter's Cathedral in Rome whose design, true to the ambitions of the Renaissance to embrace all of Western history, is orientated to the east. This anomaly thus adds several eastern-oriented Jewish temples built before the coming of Jesus (after which temples were oriented toward Jerusalem) into the Renaissance-inspired absorption of all of history into Christian renewal. That the New Testament builds upon the Old Testament is not an insignificant amendment to such symbolism.

I was struck by the incredible incongruity of the Almudena Cathedral's orientation given the problematic issues associated with the Inquisition. Surely, I thought, this canonically correct Catholic country that brought us the infamous Inquisition would not consciously misjudge the orientation of Madrid's most major cathedral.

By using scaffolding as an essentially temporal design element, my intention was to challenge the existing orientation of the cathedral by reorienting the Almudena toward the west. In so doing, the entrance to the cathedral would face downhill to the Campo del Moro, the large green open field where, early in the 16th century, Inquisition priests enunciated the expulsion from Spain of Moors, Jews and others not of the Catholic faith.

Model of urban intervention (the Almudena), Madrid, Spain, 1992

Some years later at the University of Salamanca in Spain I delivered a paper at a liturgical conference to which I had been invited that addressed that unbuilt intervention that I had been asked to design earlier in Madrid. The Spanish architectural historian Rosa Faes and I attended the Salamanca conference where coincidentally the architect of the restoration of the "mis-oriented" Almudena was also present to hear my criticism of the Almudena's orientation problem and my wonderment as to why the restoration did not take that problem into account. The proceedings were published as *Arte y Fe: Actas de Congreso de "Las Edades del Hombre"*, Salamanca, del 25 al 29 de abril de 1994; edicion a cargo de Adolfo Gonzalez Montes; Organizacion de la Universidad Pontificia de Salamanca.

*   *   *

As a by-product of my many trips to Spain, and as a result of the 1976 European motor trip I took with my children after that year's Venice Biennale where we saw our first bullfight at the Roman theatre in Arles, France, I began to frequent an increasing number of corrida. For reasons that I still do not understand, my daughter Tracy simply adored the corrida in Arles, while my son JJ and I were initially repulsed by what we misinterpreted as its brutality. It took me some time to appreciate that the bullfight is a blood ritual entirely disassociated from sport as we understand it in North America.

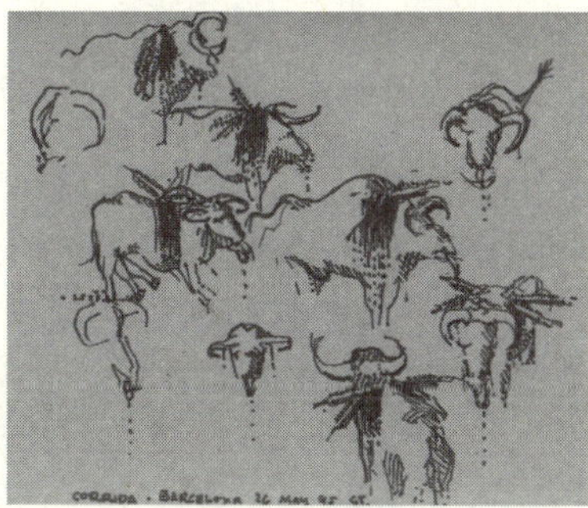

Ink-on-paper studies of *El Toro* at the *corrida*, Barcelona, Spain, 1995

Beginning with Arles and then Madrid and often in the company of the architect Raphael Moneo (an aficionado of the corrida himself) and finally 20 years later in Barcelona, I gradually became an aficionado of the blood ritual myself. My interest in the corrida coincided with my personal interpretation of the east/west orientation that is usually associated with the liturgical ritual resulting from Catholic theological antecedents.

The corrida always occurs in the afternoon. Traditionally, the bull enters the arena from the east and is momentarily blinded by the western sun. The advantage is given to the torero who is at the outset positioned in the shade of the west just beneath the box inhabited by the king of Spain. My imagination suggested that the bull was the Jew and the torero represented the Catholic priesthood that reached its apogee in the 15th-century Inquisition, or if you will, the apogee of the blood ritual inflicted on both Jew and bull.

*   *   *

I was given an opportunity to expand upon the theological implications implicit in architecture when at the end of 2000 I was invited by the Holocaust Memorial Foundation of Illinois to engage in an interview process from which an architect would likely be selected to design a major three-part cultural facility: a museum, a memorial, and an education center for their foundation commemorating an event so horrific as to beg the question. Such a building was as much about those who had expired as it was for those who survived. Working on the project might give me an opportunity of designing a space that had an unheimlich quality about it. If built, I initially thought that it could be the project of a lifetime, at least in my lifetime to that point.

Even so, I almost didn't respond to the foundation's request for an interview. I disregarded the invitation until my wife and partner, Margaret McCurry, reminded me of my frequent complaints that, "I had never built for my own kind." She finally prevailed upon me to be interviewed for this singular project. At the time, I didn't realize how ultimately complex the process would be relating to multiple client directives, particularly when the clients represented generational divisions, some of whom were Holocaust survivors and some of whom were their descendants who had other agendas. Independent exhibition program consultants brought on board years after the design procedure had begun would further complicate the process, but I indulged the interview course of action boldly, if naively.

Before the meeting, and while traveling, I produced a conceptual sketch of a tripartite building which reminded me of the two extant Jewish theologies that bracketed the time of Jesus with their corresponding orientations that represented the temple and synagogue eras and how the appalling leadership of the Third Reich was determined to not only exterminate what turned out to be six-million Jews, but to go further by dismantling the entirety of Jewish history and Jewish culture as well. As opposed to

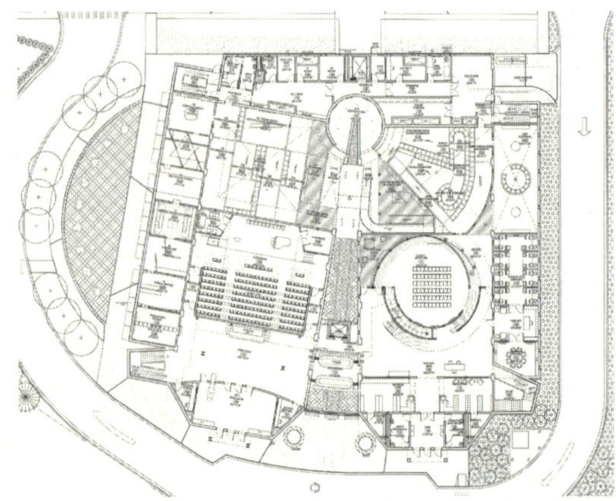

Ink-on-paper conceptual sketch for the Holocaust Museum, 2000

Computer drawing of the entrance-level floor plan of the Holocaust Museum, 2005 other architects being interviewed for the project who brought PowerPoint presentations and multiple consultants to their interviews, I brought only that quickly drawn freehand sketch with me. Within hours of the interview I was notified that the selection committee had awarded me the commission.

Nine years and two sites later, the Holocaust museum has the same original parti that was crudely expressed in that spontaneous conceptual sketch that I showed to the museum board members in 2000. The building opened to the general public in early 2009.

Not entirely unlike the earlier unbuilt Or Shalom Synagogue project, the Holocaust museum also went through some of the same trials-by-fire. The original Skokie site was rejected by the Skokie Village Council. The four gentile board members, responding to neighborhood pressure, voted against the project, while the three Jewish members voted in support of the project. The mayor of the village was disappointed in the outcome of the vote and assisted the foundation in finding another site in the same community on the other side of the same expressway and removed from any extant residential neighborhoods. As indicated at the outset of this tome, the tenacity of the architect is crucial to understanding the not-always-mysterious ways in which projects come to fruition, if indeed they ever do.

Aerial photograph of the Holocaust Museum, Skokie, Illinois, 2008

* * *

On its new site, the much expanded program for the Holocaust museum project incorporates several theoretical and/or formal concerns that have occupied my consciousness for many years embellished and developed in the context of Judaism. Unlike other Holocaust museums, I felt that overt Jewish symbolism was appropriate given the determined effort by the Third Reich to dispense with virtually the entirety of Jewish cultural history. Judaism survived its trial by fire in the Holocaust and anything less than an "in-your-face" building that posits the existential fact of its survival would have been less than what I felt was required.

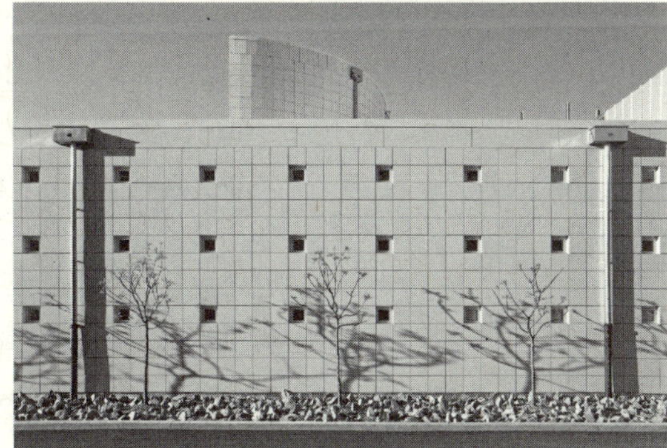

North elevation of the Holocaust Museum, Skokie, Illinois, 2009

It seemed appropriate in the context of a religion based on a sacred text that is continuously interpreted to rupture both the system itself as well as the traditionally derived forms historically produced by architects in the service of perpetuity, that is, Christianity. Furthermore, I thought it reasonable to shear such ideals by using self-similarity as a mechanism by which symmetry is disrupted, both materially and formally, but at the same time to refer back to the origins of symmetry.

I also thought that it was important to mark the presence of a visible grid as an expression of the unitary measurement of a scaffold utilizing the biblically derived cubit as a measuring tool. All this was by way of my signifying temporality for a building whose religion, Judaism, symbolized the unpredictably finite quality of life, that is, the holocaust.

The inclusion of a one-way circulation system implied that, should the observer have an epiphany after going through the museum and viewing the permanent exhibition, one would not have to retrace one's footsteps, thus seeing the exhibits in reverse in order to exit the institution. In the case of the latter commonly used strategy, the one-way directionality could also symbolize the absence of choices open to those headed for the death camps. The one-way directionality also symbolized Sheoul (Hebrew), death, not the promissory note of life after death.

That last consideration came to me after many years of giving a unique architectural problem to design students in architecture schools entitled "a circus where you entered one way as an observer and left another way as the performer." I first presented that architectural design problem to a class that I was teaching at the University of Houston, where at her final jury one of my students burned up her solution to the problem in a markedly theatrical performance where she exaggerated the epiphany possible when "transformation" is the name of the game.

The only other project that I designed that utilized the processional mechanism of "no return" was the Gulbenkian Foundation exhibition on Chicago architecture, where the logic of conservative versus liberal origins could be best explained by that technique.

*　　　*　　　*

Scaffolding per se and its repetitious measurement system which, for the purposes of this example was used to express the biblical cubit, has always fascinated me, particularly as it relates to the design of a structure in the service of Judaism, a religion in some ways defined by its implicitly pending quality even as its adherents await their own version of the Messiah. A scaffold is an ambiguous form. One is uncertain as to whether it is a sign of a building-in-the-process-of-becoming or a formerly completed structure in the process of eradication. Given that many Jews still await the coming of the Messiah, Or Shalom's reformed rabbi's cynical statement about "messianic immanence being a bankrupt concept" notwithstanding, the scaffold as a design device represents a kind of Kierkegaardian signification of that condition-of-becoming. The dimension defining the scaffold that I employed at the Holocaust museum is the secular cubit, nominally 18-inches long, a bit less than half a meter. I would submit that buildings representing the Jewish program need to employ measurement by way of stipulating "certainty" in an otherwise unresolved brief.

My interest in perceivable measurement may be seen in contradistinction to the abstract way in which some architects like Louis Kahn produced structures that are bereft of measurement thereby adding to their mystery because of the lack of scale-referentiality. My fascination with that which is apparently measurable goes to the heart of my belief that architecture ought to be a "communicable disease." Rather than obfuscating the methods by which buildings come into being, I personally find it more reasonable to communicate whatever mechanisms one uses to make the transition from concept-to-concrete such that others can participate in the process. To me that system is preferable to simply presenting the product itself and letting the cards fall where they may, thus perpetuating the myth of the architect who by distancing himself from society can lay claim to his privileged access to some kind of truth only accessible to him.

Measurement in particular in the context of a building devoted to Judaism inescapably refers back to the Torah with its many measurable descriptions from Genesis through 1 Kings, from the Ark of the Covenant to Solomon's Temple, are delimited by measurement by way of specifying presence.

In any case, when sign, symbol, site and all of the other attributes that define architecture are removed, only measurement remains. Ultimately, if "taking the measure of things" unpacks architectural mystification, that fact alone justifies its use.

There is another reason to share a building's raison d'être. By demystifying architectural concepts, one is forced to move on to new turf, lest one begins to repeat one's self-believing in the value of one's own signature. In any case, "clipping coupons" has never been a favorite pastime of mine.

*   *   *

Many factors contributed to preparing me for the Holocaust museum assignment. In addition to writing *The Architecture of Exile* in 1988 and working for more than a decade and a half on the ramifications

Ink-on-paper sketches of scenes in Israel, 2004

Ink-on-paper sketch of Abu Simbel, Egypt, 2007

Ink-on-paper sketches of scenes in Egypt, 2001

of Jewish mysticism as it relates to architecture and ethics by emending the text *Failed Attempts at Healing an Irreparable Wound*, a visit to the concentration camp in Dachau, Germany and the notorious Babi Yar in Kiev in the Ukraine had an important impact on the way that I approached the project.

In addition, studying the Jewish presence upon the land first hand did much to help enculturate me. During two separate trips to Israel separated by 27 years, I gained valuable insights into the dialectically inclined differences between tribalism and the city-state. King Solomon may have built the temple in part because of political reasons to coalesce tribal differences, but three millennia later when the Jews thought they were an innate part of the German city-state the Third Reich had other thoughts.

The second trip was particularly influential, since it occurred four years after I was commissioned by the Holocaust Foundation of Illinois to undertake the project. Among many images that I referenced as I continued to refine my design for the Holocaust museum in Skokie, Illinois were: Qumran, the site of the Dead Sea Scrolls; the Dead Sea; Masada, where on a mesa among the 1st-century ruins of the last stand of the Israelis who committed suicide rather than surrender to the Roman legions causing recruits to the Israeli army to take the oath "never again;" the ruins of Bet Shean; and the remains of the synagogue at Capernaum.

In the same spirit, during the evolution of the design of the museum I traveled to Egypt twice, which grounded me further in that portion of the biblical story as told in the Book of Exodus about the flight of the Jews from their bedeviled life under the pharaoh Ramses II to their odyssey after they crossed the Red Sea leading to their 40-year journey through the Sinai Desert as they sought an ideal referring back to the Garden of Eden. Seeing the red-haired, hawk-nosed, mummified-face of Ramses II in Cairo's Egyptian Museum and knowing that Moses was thought to be raised as his brother was as startling a glimpse through a door to the ancient world as I have ever experienced.

Since both Egyptians and Jews, among other agricultural peoples, oriented their religious rituals to the rising sun in the east, I felt compelled to return to Egypt in 2007 in order to complete my Egyptian experience by sailing up the Nile from Giza to Abu Simbel at the Sudanese border to establish the precedent of the importance of "the journey" in my own mind even though it was more than three millennia after Moses and his flock undertook their odyssey seeking a promised land.

*   *   *

With respect to my enthusiasm about the design of the Holocaust museum, my grandfather played a seminal role in influencing my attempts to bring to consciousness the importance of what he first saw unfolding in Europe at the end of the 19th-century and near the end of his life in the 1930s, but other life circumstances influenced my design as well. We each have it within us to favor one view over another, and frankly to modify those views at will. In the context that I continuously resist being defined within any given milieu, I find it useful to avoid being identified with one or another métier.

I continue to exercise my fascination with architecture as it comes into contact with the Jewish religion and the potential linkages that might result through this juxtaposition. Since architecture and the Jewish project are unexpected dialectical bedfellows, vast areas of exploration await. The overwhelming presence of measurement throughout the Torah is only the beginning of an architectural debate that promises to enliven those involved in both disciplines.

In particular, my long flirtation with Judaism as an ethical operation distinct from its religious dimension remains an important factor as I work through certain unresolved issues that continue to intrigue or plague me depending on one's point of view.

During the time frame that I began to define linkages between Judaism and architecture, I was interviewed to design a synagogue in Minneapolis. The congregation's chief rabbi criticized me by saying that, while he had read my book *The Architecture of Exile* with interest, Judaism was not simply an intellectual pursuit.

Uninhabitable void, Holocaust Museum, Skokie, Illinois, 2009

Of course, the rabbi was correct in his assessment of my academic enthusiasms, perhaps even eagerness to compartmentalize a subject like Judaism more normally practiced than intellectually categorized. I am reminded of Socrates who had conveyed the importance of "habituation" to his students in the context of ethics. His point was that for an ethical position to be an intrinsic part of one's being, one needed to make it a significant part of one's life by practicing it regularly, which is to say making it a habit. In all events, I remain determined to coax out whatever relationship that exists between these two perhaps disparate callings, architectural philosophy and Jewish theology.

Most importantly, my increasing concern with ineffability led me to unpack the two orientations extant in Judaism that I then concretized in the Holocaust museum. The dark building symbolized the descent into the darkness of the museum while the ascent into the light represented the hope implicit in education. When those two parallelepipeds were hinged at the western end of the structure, they revealed a space that could be construed to be ineffable in that it was uninhabitable by the living. Indeed, it was meant for those who didn't survive the Holocaust and became inexplicable in the way that the unheimlich created a feeling of the uncanny.

The uninhabitable void of the Holocaust museum is its most important space precisely because it is inaccessible. The desire to penetrate that which is impenetrable is overwhelming, but the cleavage between the two rectangular parallelepipeds is not for the living; it is dedicated to those who have perished.

4

ETHICS IN THE CONTEXT
OF SOCIAL CAUSE

Fundamentally, architecture is nothing if not principled. Simply by placing something as consequential as a building on the planet where nothing existed before that time requires an optimistic determinant, without which it is extremely difficult to practice architecture with any deeply felt dedication. To bring a building into existence requires an enormous act of will. One can only hope that architecture as a discipline is informed by ethics. Even though traditions within architecture appear to reify ethical practice, architectural students are not always exposed to value-qua-value in the context of design. They spend a great deal of their time in school developing the many technical and aesthetic skills required to practice the profession competently, skills that it has been determined will make them employable.

*   *   *

The American Institute of Architect's (AIA) latest version of its much watered-down ethics clause results in part from a class-action lawsuit initiated by the U.S. Department of Justice and settled by both parties in 1990. The resulting consent decree stipulates that all architects who join the AIA are required to abide by the decree that enjoins the AIA from having an ethics clause that would basically impede architects from engaging in what the U.S. Department of Justice referred to as price fixing: free enterprise.

At that time, the Chicago Chapter of the AIA had issued a study outlining suggested fees for architectural services in an attempt to dissuade firms from undercutting each other's fees in an effort to secure clients thereby jeopardizing the already low salaries paid to employees or that put some firms out of business. In the revised version, the AIA stipulates that architects are now allowed to "undercut fees, advertise and supplant others" without notifying them or assuring that those displaced architects have been recompensed for services they have provided. This is not the discipline that I signed on for more than a half-century ago. Times seem to have changed in ways that I neither anticipated nor believe in, or frankly, even appreciate.

In 2002, I revised a paper I had written earlier entitled, "The Case for an Architectural Hippocratic Oath," originally published in the *Harvard Architectural Review* in 1992 and formerly entitled "Towards an Architectural Hippocratic Oath." In that paper, I delineated those constituent features that outline the case for an ethical practice.

It may be that because architectural practice has never had its own Hippocratic Oath, other less-than-laudatory free-market-based capitalistically driven forces have arisen that now tend to diminish the discipline such as marketing and branding, or in a word commercialization. This trend seems to have become the 21st century's sign of the displacement of ethical responsibility as an ethos that in prior eras in the United States played an important role in the architect's behavior with respect to professional standing.

I once observed that "the practice of law is a perversion of the study of law" by which I meant that the noble nature of the study of law is sometimes perverted by the exigencies of practice. If the professional practice of architecture and the expediency connected with commercialization continue to undermine the ethical determinant that I so fervently believe needs to be the basis of architecture, soon one may be able to suggest that "the practice of architecture is a perversion of the study of architecture."

Practice seems to be the inexorable result of study and as such, suggests that study is secondary to practice. I prefer to think that practice needs to be more like study. The practice of architecture would be well served to seek the high road inherent to study rather than suggesting that a straight line is the shortest distance between two commercially defined points, or that architecture is solely about actualizing concepts that might be better served unbuilt.

*   *   *

Throughout the 20th century, certain precedents other than commercialization have sometimes also had a negative impact upon the field of architecture. For example, Fikret Yegul's controversial book *Gentlemen of Instinct and Breeding* published in 1991 describing the early years of the American Academy in Rome, points out that racial, religious and gender discriminatory practices were imposed on those architects who applied for admission to that noble institution. Arguably, one can make the case that restrictions per se were in the air during the time span of Yegul's argument, but one can also make an equal and opposite case that during virtually all of the first half of the 20th century the American Academy in Rome did precious little to challenge a code of behavior that was anything but egalitarian.

It was largely a result of the democratization following the Second World War with its enormously varied cadre of returning veterans that Catholics, Jews, Muslims, women, people of color and Native Americans began to penetrate the architectural field at all levels of the discipline. They were to become academics, practitioners, theoreticians and critics. Those same culturally, racially and ethnically diverse veterans had an equally substantial impact upon cultural institutions such as museums, which began to replace entertainment with education as they felt the need to redress certain biases that, until that time had been de rigueur throughout many aspects of American life.

*   *   *

Beginning in the late 1970s and newly remarried to my third wife, my own life began to change, and with it the nature of my architectural practice had also become modified. It struck me increasingly that architecture was perhaps more about working for those in society who cannot always afford architects than it was about addressing design problems for those who can afford to hire an architect. Thus, and clearly belatedly, I began to understand the importance of giving back to society.

My earlier projects in the Chicago region that perhaps qualified as addressing problems which were need based such as the Illinois Regional Library for the Blind, The Anti-Cruelty Society, a summer camp for disabled Boy Scouts and another camp for abused children administered by United Charities all impacted the way I perceived architectural practice. So did a later project for preschool children on an at-risk site adjacent to public housing as well as a project designed to provide legal, medical and sociological support for sexually abused children. These projects gratified me in ways that formalistically informed architecture-for-its-own-sake didn't. This attitude doesn't make other kinds of architecture that could be construed as exclusivist necessarily less meaningful than those I gradually chose to expend energy on, just different.

Designing the Illinois Regional Library for the Blind represented a watershed in my approach to designing for those in need. While I produced a building whose focus was to respond to sightless persons and their idiosyncratic requirements, such as both linear and fixed elements and the use of color that could be decoded by near-blind persons, it was the inexplicable concrete wall with its very long structurally unsupported horizontal slit window that symbolized the extra-rational condition of blindness that brought home the importance of aesthetics for its blind constituents.

Even so, I had difficulty in persuading sightless users about the value of designing a building that was barrier-free let alone one that might appeal to their aesthetic sensibilities. Their argument was that they spend most of their time in buildings that are laden with barriers and that a building designed specifically for them that accommodated their needs would reduce their ability to cope with the many barriers that they normally face. Though I prevailed in the end, sometimes you can't win for losing.

The projects listed above, when combined with my several Chicago-based social housing projects and the vocational training schools in Bangladesh all conspired to substantially change my views about architectural practice and its potential democratization. The project for disabled Boy Scouts is a case in point. Dating

Adirondack cabin at Camp
Hoover, Yorkville, Illinois, 1985

back to the Norman Rockwell *Saturday Evening Post* magazine cover depicting the "able-bodied scout," the Boy Scouts of America has been inadvertently portrayed as a kind of paramilitary organization whose purported mission was, nonetheless, to develop "character."

In contradistinction to that image, the Chicago Council of the Boy Scouts initiated a joint program aimed at servicing disabled Boy Scouts by cross mentoring them with able-bodied Cub Scouts at the same time. The project was a breakthrough of some magnitude in which I was fortunate enough to be involved. At the summer camp run by the Chicago Council, the disabled older Boy Scouts were to be paired with able-bodied Cub Scouts. The intention was to have the older disabled boys sleep in the lower bunks, while the able-bodied younger boys would sleep in the upper bunks; thus, cross mentoring of various kinds could emerge from such a unique juxtaposition. The program was nothing if not pioneering.

From a perceptual point of view, the Adirondack-like cabins that I designed for the Scouts seemed to me to also symbolize the hybridized quality that I felt stood for the unique multivalent population of the United States, to say nothing of the neo-militaristic quality that the cabins represented for me. European stylistic precedent combined with agrarian functionalism has always seemed to me to represent a significant architectural melding of our mixed heritage.

My latent attitude about this subject was brought to consciousness through the utilization of the signs and symbols that Margaret and I employed in the design of our own rudimentary rural Michigan weekend house. Its corrugated steel clad form can be "read" as a metaphor for either a basilica or a barn with an attached baptistery-cum-grainery. The doubling of architectural language bridging the theological with the rural is well within the understanding of American hybridization.

I continued this research into the American lingua franca with two other projects, the first of which had an unusual human dimension that informed its program as well. My clients were retired educators for whom I had previously designed a house in the Chicago suburbs. Though never built, this project for a Christmas tree farm in Kenosha, Wisconsin was situated in a prairie-like setting common to the Midwest on property that the two university professors owned. The tree portions of the farm were to have been bracketed by the public face of the project, a greenhouse sales center that flanked a parking lot adjacent to the road racing by the site. The private face consisted of two extremely modest retirement houses that flanked a lap pool with the parents' house at one end and their disabled daughter's on the other. Fir trees were to be grown in four distinct precincts based on the age of the trees, and the rural quality common to the two earlier projects was also present in this agrarian complex.

Ink-on-paper drawing of Christmas tree farm, Wisconsin, 1993

The second indigenous rural American project was the vernacular design of a number of outbuildings for a Chicago-based lawyer and gentleman farmer whose 1,000-acre farm near Three Oaks on the Michigan/Indiana border was leased to locals for the management of crop production, but was used by his family as a weekend getaway. He had excavated two large ponds for use as swimming-cum-fishing holes and saw the outbuildings as useful components that would complement this otherwise agrarian scene.

Boathouses, Three Oaks, Michigan, 1996

*   *   *

At an urban scale, two projects came into the office at the turn of the 21st century that further reinforced my feelings that architecture in the context of social cause has the utmost reward system. Designing merely to embellish one's formal vocabulary for use in the development of an interpretation of architectural language had for me become a limited, ultimately unsatisfactory pursuit when addressed in isolation and detached from giving back to the community.

The first of these was a preschool project that was to be built in a challenging urban neighborhood that, at the time, was a refuge for drug dealers, "gang-bangers" and drive-by shooters. The foundation committed to changing the calculus of children's education stood in the way of the metropolitan inertia that diminished life in that part of Chicago.

The Ounce of Prevention Fund was the brainchild of Irving Harris, one of the most philanthropically inclined early education advocates for disadvantaged African-American children in the United States. He had developed a number of preschool facilities and he was interested in doing a "flagship" facility that

Educare, Chicago, Illinois, 1997

through its design as well as its program might act as a paradigm for future facilities as well as protecting young children from the temptations of gangs and drugs. Educare was to become a prototype for a physical plant that accomplished his ambitions, and I was fortunate in being selected to design that facility. The building was to be located between a neighborhood grammar school and the historically well-known DuSable High School.

We proposed to build the facility around an open atrium that could act as an outdoor play space and that the building itself would insulate the children from the several deleterious influences posed by the particular neighborhood in which it was to exist. The façades of Educare were designed to be unthreatening to the neighborhood so that the community would embrace the new facility as a contextual extension of the residential accommodations abutting it.

Interestingly, many of the neighboring public-housing high-rises were demolished shortly after the Educare complex was completed. Even so, bullet-resistive fences were erected to deter drive-by shootings and the open atrium served its constituents well by providing protected spaces where the children could play in all temperatures, and where their teachers could rest in the shade as their charges went about their business.

*   *   *

The second urban project that also fell under the rubric of architecture in the context of social cause was commissioned by the city of Chicago to provide facilities to house a variety of advocates who, working through the court system, were available to represent sexually abused children. The Chicago Children's Advocacy Center was designed to address the many programmatic issues facing the disparately derived

Chicago Children's Advocacy Center, 2001

constituency groups who the building was to house. The Chicago Police Department, the State's Attorney's Office, a medical team from Cook

County Hospital, the Department of Children and Family Services and social workers were to be housed together to contend with the abused children within the legal system that existed to serve them.

I attended 18 months of meetings between members of the various city departments who had been assigned to develop an interface between them. The law stipulates that methods of interdepartmental relations be discrete one from the other so as to ensure fairness for all parties once the justice system takes over. The plan evolved naturally about two interior courtyards. Pedestrian and vehicular entrances that axially opposed each other were adjacent to the two courtyards.

My purpose in utilizing crayon-colored glazed brick, installing colorful awnings, applying window boxes with flowering plants and playfully skewing windows was to break down the feeling of an institutional structure to make the building welcoming to sexually abused children. Considering the traumatized state of the very young innocent user, it seemed to me that the architecture must not be further intimidating. That mentality caused me to create courtyards to protect the user from intimidation by the alleged perpetrator who I was informed sometimes follows the abused child and his/her family to the center.

*   *   *

Like the cloisters that evolved in 10th-century Cistercian monasteries, I began introducing contained spaces located in buildings that I designed that were located in an urban environment to protect a building's constituents. It struck me that the benefits of axis mundi that one finds at the center of theologically informed cloisters often transcends buildings that may be religiously driven but are intrinsically important themselves, particularly to at-risk users. In all events, cities sometimes give the impression of alienation and the use of cloisters ameliorates such anxieties by the intimacy and security implied by protected spaces.

An additional benefit of cloistered spaces was a potential aura that might come about as a product of the architecture that defined such spaces. That aura, or ineffability, was something I had been searching for as my work evolved. Certainly, there was a precedent for such spaces, but that precedent had generally been situated in theologically derived buildings.

*   *   *

Giving back to the community through the vehicle of architectural practice is one way of coming to grips with a social dimension that I consider necessary to fulfill an architect's required ethical dimension. Another mechanism is to find ways to break down barriers between theory and practice, in order to challenge those involved in the various aspects related to defining the built environment to accomplish something

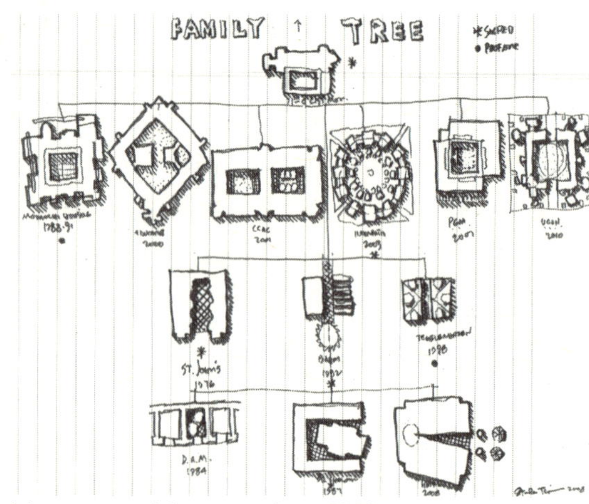

Ink-on-paper cloister genealogy, 2008

more than solely actualizing it, or contradistinctively by exclusively theorizing about it.

In 1980, I was persuaded to return to the University of Illinois at Chicago (UIC) to direct their one-year post-professional graduate program in architecture. Referred to as Option One, the program was directed at those who already had a first professional degree in architecture. I had been absent from a full-time educational position for virtually a decade, so I was intellectually eager to be involved in a more theoretical milieu in which I planned to test my mettle over and against the "anything goes" architecture school that I had resigned from 10 years earlier.

By way of initiating a dialogue within the school between theory and practice, I invited some of the more visible architectural luminaries of the day from around the country to UIC as visiting critics to teach alongside me. Heather Cass, Peter Eisenman, Eric Owen Moss, George Ranalli, Robert A.M. Stern, Susana Torre and Lauretta Vinciarelli were among those who brought a refreshing outlook to the school. Many of the teacher-practitioner folks that came to UIC had strong theories underpinning their architectural production that I felt would be useful in an academic environment that had, up until that time, been situated firmly within the Midwestern pragmatism for which our region is so well known.

Another way that I challenged the hyperpragmatism for which the school was noted was to sponsor John Hejduk and Peter Eisenman to receive honorary doctorate degrees. Their honors also suggested that theory as well as practice underpinned the architectural profession contrary to the way by which practicing architects had traditionally been selected to teach at UIC.

Within the broadest possible stylistic range, the presence of nationally respected architects and academics had a salutary effect on the students. The visitors also opened up debate between disparate factions present both within the profession in the city of Chicago as well as at UIC. Other, younger designers and intellects came to critique the sometime mindless approach to architectural education that so many design schools between the mountain ranges favored. Doug Garofalo, Catherine Ingraham, Mark Linder,

Greg Lynn, Eva Maddox, Steve Perrella, Mark Rakatansky, Maria Smithburg, Bob Somol and Lily Zand brought both energy and a much needed dose of modernist architectural sensibilities to the tenured faculty, most of the rest of whom were, simplistically speaking, prosaically inclined practitioners. These cutting-edge individuals could also mix it up on juries and in debates with others I had hired like Thomas Gordon Smith, whose entire career represented a slavishly unswerving dedication to the traditions and values of the classical language of architecture.

Nonetheless, my commitment to diversity notwithstanding, at some point the obligations of leadership could be construed to demand that a school represent something directional as opposed to continuing the laissez faire attitudes prevalent at the institution for the many years of its existence. With that in mind, I organized and curated an exhibition at the UIC Gallery entitled Ten Untenured Faculty. The exhibition represented theory, history and criticism over and against pragmatically derived architecture, interiors and landscape. The implications of physical adjacency between exhibitors required that the participants consciously contend with each other rather than simply presenting their own views independently. It was an extraordinary exhibition that also represented the state of the art of architecture in the United States generally and Chicago specifically at the time.

Most of the participants have gone on to illustrious careers in their individual specialties, but at the time, the show created quite a stir among those who had risen to power within the authority given to tenured faculty at the school. Given the entrenched position of the tenured faculty, that disquieting exhibition represented a threat to their inertia. In some ways, the exposition also represented the beginning of the end of my usefulness at the helm at UIC's architecture school.

At the time it seemed that UIC's architecture school was making a mark of some sort. To further that momentum, and on hearing that Notre Dame was seeking a new architecture dean, I contacted the provost of that illustrious institution and recommended Thomas Gordon Smith for the appointment. I told the provost that "classicism" needed a home and that a university looking to Hellenic Christian precedents seemed to me as an ideal repository of such a direction. Smith seemed to be an ideal candidate for a position thusly defined and without further ado he was offered the deanship at Notre Dame's architecture school, and promptly accepted.

*   *   *

As the designer Bruce Mau has said, "there are no boundaries any longer." I would expand on that comment by suggesting that whatever barriers still exist between disciplines as well as between the academy and practice need to be removed in order to cohesively move forward into our emerging multivalent future in the 21st century.

In any case, the design professional of the foreseeable future will be one who is equipped to operate in a variety of circumstances well beyond the normative one with which we are familiar, that is to say monolithic practice devoted to the making of objects alone. Beyond that, new expressions denoting the process by which structures evolve will play an increasingly important role for those practicing and/or teaching well into this new century. Equally important, ethical practice denoted by two subsets, sustainability and social cause, may well define the immanence implicit in a multivalent practice.

*   *   *

I encouraged Gianfranco Monacelli, who at the time was directing Rizzoli's American publishing activities, to simultaneously publish two different Chicago architectural journals that I felt would add significantly to the energy of Chicago's architectural community such that Chicago architects as well as architectural educators, theorists and critics could more comprehensively address the major issues of the day. The first journal that Rizzoli published was the *Chicago Architectural Club Journal*, which recorded the proceedings of the newly reinstated Chicago Architecture Club that had initially re-emerged due to the efforts of the Chicago Seven. The second journal to appear was *Threshold*, the journal of the School of Architecture at UIC. First published by Rizzoli in 1982, *Threshold* continued to be published for a decade, with a total of five issues finding their way into print. It concluded with Threshold 5/6 published by Rizzoli in 1991. The *Chicago Architectural Club Journal* was first published by Rizzoli, Chicago in 1981, and thereafter four successive journals appeared.

To some degree, these two Chicago-based journals helped to open up the Chicago architectural community to much needed dialogue. Advocates of the overly well-known second Chicago School, predominately represented by Mies van der Rohe-trained descendents had persuaded like-minded architectural historians that nothing of note was happening in the city to transcend the canonically accepted view that premiated "structure" and "construction." But by the 1980s, revisionism was in the air and had become an important influence among Chicago's architectural and urbanistically committed newest constituents.

In any case, architecture in the 20th century had always been concerned with "the city." I was reminded of that fact when I was a visiting critic to the architecture school at Iowa State University in the late 1980s. On one particular visit, I expressed interest in the opinion of the architecture student assigned to pick me up at the Des Moines airport. As we drove to Ames through the flatlands of Iowa, we discussed where he thought architecture was best practiced, vis-à-vis the city or the country. Citing Iowa State's urban program in Des Moines, he unequivocally responded that "architecture is about cities," and I realized that while architecture is in fact of and about cities, the kind of city typologically speaking is as yet undefined in the 21st century.

Is it the formal city as some of us have always understood it replete with census tracts dictating rules, conscription and taxes, that is, order, or is it the informally inclined favela-like city with its squatter constituents, who often override rules with barter and trading services? Stewart Brand, along with other futurists, reflects about these issues continuously, reminding us about the several possible ways that we all relate to each other in spite of our differences.

When I asked the Iowa State student about his thoughts relating to the needs of the city, I had in mind something other than the response he gave me about the value of urbanism. Because I was writing *The Architecture of Exile* while teaching at Ames, the concept of the Garden of Eden was on my mind. I felt that the "new Eden" was in fact the city of man, and that ideality was most clearly represented in the urban environment, not in a return to nature.

*　　*　　*

In 1984, Yale's architecture school conducted a search for its position of dean that had become available. Three Yale architecture-school alumni—Tom Beeby, then UIC's architecture school director, my old friend the New York architect Charles Gwathmey and I—were selected as the short-listed candidates, and each of us in turn was interviewed by the Yale School of Architecture faculty, the current student body and, significantly, by Yale's then president A Bartlett Giamatti, a distinguished Renaissance scholar who, after his tenure as Yale's CEO, subsequently became commissioner of Major League Baseball, a position that he held until his untimely death.

During my particular interview with him, President Giamatti asked me to explain who I was innately in one word. Without any thought and in a nanosecond I blurted out, "Action." Later, I suggested to Tom Beeby that when he was asked the same question, he probably waited a half minute or so before carefully replying, "Contemplation." Typically, Beeby smiled knowingly but didn't react directly; needless to say,

Beeby got the post.

In 1985, with Beeby now safely ensconced in New Haven, I was asked to succeed him as director of UIC's architecture school. Before he left for New Haven, Beeby advised me to treat the position of UIC director as if I was a new cardinal in the Vatican's College of Cardinals, observing and listening, but not taking action during my first five-year term. Of course, inaction would have been the last thing in the world I would have favored, given my short attention span. I have always believed that the worst that can happen to those who are proactive are acts of commission as a result of wrong-headed decision making, while deferring issues can result in acts of omission. I was far more interested in achievement versus observation of the existing academic environment whose tradition at UIC, in any case, was grounded thoroughly in a laissez faire state, where anything and everything was allowed to co-exist together bereft of any particular direction that might have jeopardized that condition.

I accepted the UIC directorship with the understanding that I would attempt to enhance the general perception about the architecture school. Actually, I aspired to take the school in a direction that I hoped would position it internationally. One of my initial activities as director was to submit to an exhaustive interview with the journal of the Association of Collegiate Schools of Architecture (ACSA). In that 1985 discussion, I outlined the value of juxtaposing architectural history, theory and criticism with practice as a way of melding presumably paradoxical activities thus presaging my eight-year tenure as director of the School of Architecture at UIC. I also layed down a kind of gauntlet to the faculty by putting all concerned on notice of my theoretically inclined aspirations for a school such as UIC where architectural practice was the be-all and end-all of the entirety of its history and which in all events, like much of the Midwest's reputation, was pragmatically driven.

In any case, the tradition of an architecture school situated in Chicago was anything but intellectually inclined; witness for example IIT. In part, Mies van der Rohe's success in Chicago both from an architectural education point of view as well as in his architectural practice was that Chicago was in some ways responsive to Mies's perhaps apocryphal dictum "build, don't talk." Chicagoans are routinely apprehensive about intellectually informed verbiage, particularly in a field like architecture that is driven by actualization more than it is by hypothesizing.

The immediate challenge that I faced at UIC was its accreditation dilemma. Just prior to my becoming director, an accrediting team from the ACSA had, for the first time in the school's history, put it on probation by recommending an abbreviated term thus diminishing the program already in place. I felt it important that, with a new director in place, it was crucial that we started out with a clean slate with full accreditation in hand. I asked for and received an appeal hearing, told the review board of my plans to introduce theory to the program and, within a short time, acquired the full accreditation for the school.

Shortly thereafter I organized an international conference at UIC, that we referred to as the Chicago Tapes, inviting a gathering of internationally well-known practicing architects, theoreticians and educators from Europe, Asia and the United States to Jane Adams' famous Hull House where each participant was to present a recent project. The assembled body was then to respond critically to the individual projects that had been presented by their colleagues.

The conference was covered by Rizzoli editor Gianfranco Monacelli and the ensuing book that came about through audio taping and its subsequent transcription was then published by Rizzoli International. It followed on the heels of a similar conference with many of the same architects that had been convened earlier at the University of Virginia. These kinds of events seemed to suggest that UIC was quickly becoming a player in things intellectual as well as beginning to be less mired solely in the well-known "Midwest pragmatism."

While the Chicago Tapes was stimulating from several points of view, the gathering was not nearly as provocative as was the preceding event that Peter Eisenman organized and held at the University of Virginia. In Charlottesville, for instance, the classically inclined architect Leon Krier appeared to insult the modernist Tadao Ando by unhurriedly and continuously clapping his hands together after Ando presented his Osaka urban courthouse residence. Krier exacerbated things further when he stated rudely that, "while the world burned, these kinds of projects still come about!" As late as December, 2008 Leon was still attacking modern architecture for acts of self-involved, formalistically driven shape making.

At a reception that evening at Dean Jacque Robertson's pavilion, Leon was playing Bach under the flickering light of a Liberace-like candelabra ablaze and sitting on the piano. Within earshot, I wondered aloud whether Leon knew that in an earlier life in Japan, Ando had been a proficient professional pugilist. It gratified me to hear Leon begin to miss notes as he continued to pound away at his little Bach fugue.

*   *   *

Almost immediately after the conclusion of the Chicago Tapes conference one of its presenters, Michael Graves, recommended that I give a major lecture in Osaka, Japan. While on that trip, Margaret and I were escorted by Tadao Ando and his wife to the northern provinces of Honshu to see what for Ando was an especially important ancient Buddhist monastery. He insisted that we arrive shortly before sunset to get the full effect of what we were about to see, driving his Porsche accordingly at speeds of up to 100 miles per hour to guarantee that we would make that deadline, the white-knuckled apprehension of his passengers notwithstanding.

Compared to the well-maintained Shinto temples we had visited in Osaka, we arrived at a humble wood structure placed seemingly without rhyme or reason in the wild landscape of northern Honshu. We had to wait awhile in the not-quite-manicured courtyard, since the caretaker would only allow us inside moments before the sun went down in the west. Noting that the timber structure that housed its larger-than-life Buddha and many selected details were stained blood red, we were nonetheless overwhelmed when the shutters were thrust open to see that the golden light of the setting sun amazingly appeared to make the entire building's interiors along with the resident Buddha go up in flames! As we sat quietly cross legged on the floor contemplating this phenomenon, I began to understand the source of the otherwise inexplicable emotional component of Tadao Ando's powerful architecture.

I felt that I was in the presence of a method-of-making akin to the mysterious (to a Western-educated mind) wabi-sabi and nothing that was intellectually rationalized could describe what motivated Ando in his quest for an innate poetic sensibility more than what that ancient monastery offered us that late afternoon. I remembered Leon Krier's thoughtless comment in Charlottesville and wondered what Leon might say about the phenomenon that we had witnessed at sunset in the rural provinces of northern Honshu.

It also gave me pause to contemplate the emotive component prevalent in much Asian architecture versus the intellectual way that architecture often comes about in Western culture. Finally, I began to see what Frank Lloyd Wright understood about Japanese architecture that led to his well-known symbiotic rapport with that civilization. From an architectural point of view, the Asian use of syntax as a replacement of the Western use of symbolism to define buildings beyond simply expressing their functional resolution was a revelation to me personally.

*   *   *

On the subject of ethics in the context of architecture, it seems to me that the Japanese commitment to making buildings well is fundamental to such a pursuit. My early experiences with George Fred Keck, Mies van der Rohe and Paul Rudolph all influenced me to give equal credence to structure and construction together with aesthetics and use. This architectural balance was also present in the Japanese psyche. Thus, my antagonism to the successive leadership at Yale's architecture school after Paul Rudolph, that is to say Charles Moore's stewardship and Venturi's participation in it, is in part based on Moore's advocating the use of "2 by 4 wood framing and drywall" as a means of building and how Moore's successors tried to give credibility to methods of construction that reflected in reality that which engendered impermanence. I would propose that the unremitting replacement of buildings constructed such that they deteriorate prematurely is anything but ethical.

I believe that building well is not only the best revenge but an ethical statement in and of itself. My own fascination with the temporal implications of architectural theory in the context of Jewish theological precedent notwithstanding, to encourage or to tacitly support physical deterioration as the inevitable result of a particular construction type that is known to not last very long is, to my way of thinking, an indefensible proposition. I would suggest that building substantially is not only the result of an ethical understanding of architecture, but ultimately, is to be expected and indeed to be demanded from a responsible architect.

One of the most potent examples that clarify the dialectic between permanence and its deteriorating counterpart is to give an account of Arata Isozaki's own story of growing up near the end of the Second World War in Oita on Japan's southern island of Kyushu. He recounted this episode of his youth while he once delivered a lecture at the Graham Foundation in Chicago.

According to Isozaki, every day during the waning moments of the Second World War he and his 14-year-old classmates would play outside after school until the air raid siren warned them to scurry to bomb shelters. When the all-clear was sounded they went back outside to resume their play. The day that the atomic bomb was dropped on Nagasaki, when the all-clear was sounded and the children came out of their bomb shelter, they saw nothing but "blue sky and silence," a phrase that came to resonate with artists of Isozaki's generation. As a result, Arata Isozaki was determined that he would build so utterly enduringly both symbolically and actually that the buildings that he designed would never be subjected to erasure in an uncertain future.

Certainly, the Zen-like philosophy of the Asian aesthetic known as wabi-sabi, suggests that we consider an alternative understanding about nature and the impact that it has on human beings. While buildings, like bodies, begin to die at birth (compellingly stated in the late 1980s by the former architectural educator John Whiteman while he was director of the Skidmore, Owings & Merrill Foundation in Chicago), I am drawn to what the title of my forthcoming book stipulates, *Failed Attempts to Heal an Irreparable Wound*. I am personally committed to a proactive posture that gives credence to the attempts to overcome inertia by struggling to build as a way of making a difference, even as one comes to realize that the struggle to heal is a losing proposition.

*　　*　　*

But it's the topic of giving back to society on which I wish to focus. Two models in which I've been involved recently address this subject, which are: 1) an alternative design educational model, and; 2) the fundamental nature of architectural practice itself for those most in need of it.

"Giving back to society" was not something that I consciously thought about when conceiving Archeworks as an alternative design educational model but the opportunity came about all at once as a result of conversations that I had with the architect-professor Douglas Garofalo, the multifaceted designer Eva Maddox and the theoretically inclined academic Robert Somol. The four of us considered the limitations as we understood them in conventional design education that hindered lessons learned in architecture schools concerning design in the context of social cause from finding its way into practice. And yet we understood that architectural projects for those most in need of them have been in abundance in design schools for a very long time indeed. Was there perhaps some synapse that we couldn't otherwise explain?

I thought that the time might be right to introduce a different educational model onto the architecture-design educational scene that: 1) eliminated tenure; 2) did away with prerequisites; 3) shunted accreditation aside entirely, and; 4) attempted to break down barriers between design disciplines as well as between the academy and practice. Given that those four factors seemed to be missing in conventional design institutions, we believed that the absence of such essential elements seemed to stultify architecture schools as we understood them. Of course, all of the above could transpire as a product of inventing from scratch a design institution whose premise might come about in the context of addressing design for those most in need of it. After some hand wringing and still filled with some doubt about the whole endeavor, I persuaded Eva Maddox to join me in this adventuresome enterprise.

"Breaking down barriers" had already been introduced in my practice from a formal point of view reflected by my interest in disintegration. As I saw it, the challenges facing architecture schools represented another opportunity to deconstruct for the purpose of analysis without yet coming to grips with the value of reconstructing that which we had taken apart.

Something must have been in the air in the spring of 1993 because I found that I too had been deconstructed when I was asked to resign my position as director of the School of Architecture at UIC by the recently appointed new dean of the College of Architecture and the Arts who was responding to a tenured faculty who felt threatened by the young Turks on the faculty who I wholeheartedly supported. I refused to resign having just installed a number of determined, theoretically inclined younger academics and cutting-edge practicing architects. I felt that my willing departure would leave them vulnerable to a tenured faculty determined to return to their former laissez faire state by eliminating whatever and/or whoever stood in their way.

I told the dean that she would have to terminate me since I couldn't just leave my recent hires to the vicissitudes of the already discontented tenured faculty that opposed my attempts to either "shape them up or ship them out." Thus challenged, the new dean accommodated my concerns by firing me from my role as director of the school.

I promptly resigned my tenured professorship at the university (again) by negotiating a retirement contract with the school at arm's length. This act made it possible for Maddox and me primarily, and Garofalo and Somol secondarily, to initiate mechanisms for the implementation of what would eventually become Archeworks. We held strategic planning meetings, invited people sympathetic to our ideas about architectural education to cocktail parties to discuss our pedagogical aspirations and generally stroked ourselves into a euphoric state so as to move forward along this adventurous path towards an as yet ill-defined, unconventional educational goal.

Seventy-five years earlier, another multidisciplinary institution was founded in Germany to overcome a somewhat analogous academic stultifying condition. It was called the Bauhaus. While the economic conditions of 1993 in the United States certainly were not identical to post-First-World-War Germany, there was more than a passing self-similarity between the two epochs and their stressed economies. The well-documented economic recession in Germany after the Great War weakened the Weimar government's ability to adequately contend with the problems of unemployment, and while the major economic recession of 1993 in the United States certainly was not nearly of the same magnitude, it nonetheless had an appreciably deleterious impact on the furniture industry whose success depended upon construction in the areas of commercial and institutional architecture and design.

Coincidentally, Eva Maddox and I were made acutely aware of this ominous state of affairs as a result of our two separate firms working jointly on an urban-design project conceiving street furniture, lighting and other amenities in the central business district of Muskegon, Michigan. That joint project coincided precisely with the economic recession. Many American contract-furniture companies located in the western Michigan region, such as Haworth, Herman Miller and Steelcase were suffering from a substantial reduction in office-building construction which in turn severely impacted their individual furniture manufacturing operations.

On one particular trip to Muskegon, Scott Devon, one of the Muskegon city fathers with whom we had been working, suggested in casual conversation that what was really needed in that micro-region to help accelerate recovery from their economic woes was to organize a Bauhaus-inspired school. He suggested, and we hastily agreed, that such an institution could discover and perhaps even deliver re-purposed products in unanticipated arenas, such as, for example, prefabricated affordable housing, health care delivery systems and alternative solutions to homelessness.

While it didn't make any sense for Eva and me to leave our Chicago practices, let alone our spouses, in order to start an adventure such as the one that Scott proposed 125 miles from Chicago, it was clear to me that something of consequence had been discussed. Almost immediately I made the rash decision to start an alternative design school in Chicago and, after some serious tooth gnashing, persuaded Eva to join me in this unpredictable adventure. Stepping back from the emotionally enthusiastic outburst that led to our impetuous move, we were well aware that while conventional architectural education had, on more than a few occasions, encouraged students to design within the context of social cause, it seemed to us that precious few of those concepts ever found their way into practice.

It was just that simple that Archeworks came into being. We never stopped to think about what roadblocks might stand in the way of our idea for a new and ultimately very different design school from the ones with which we were familiar. But, as the pop song suggests, "Fools rush in where angels fear to tread." In hindsight, it seems somehow fitting that we began Archeworks on a wish and a whim, for that is precisely the stuff of dreams that has always informed the spontaneous side of architecture and design, thus grounding it in optimism; that is, entrepreneurship.

It required a year of planning before classes could actually begin, but we used that hiatus fruitfully by designing and distributing catalogs and posters to architecture and design schools worldwide as we concurrently began generating focused thoughts about what might be the nature of our mission. We consulted an attorney, who prepared us to legally constitute Archeworks as a public charitable trust—in IRS terms, a 501 [c] 3—and assembled a manageably small board of trustees in support of our cause. Significantly, we were given a basement space rent-free in a loft building on Chicago's Near South Side across the street from H. H. Richardson's historically famous Glessner House.

*   *   *

In the fall of 1993, Fred Koetter, the dean of the Yale School of Architecture, appointed me to a semester-long visiting chaired professorship, which I then used as a "dry run" for Archeworks. During that semester, I observed that Yale architecture students who had signed up for the studio seemed to be enthusiastically responsive to the particular social issues that informed their work. They produced astonishingly innovative solutions to problems that they themselves conceived. All of their projects were in the context of social cause and many of them were situated in the vicinity of New Haven. Garofalo, Maddox and Somol joined me in New Haven for the final review, thus giving each of them the opportunity to see what an Archeworks jury might look like.

In addition to the Yale architectural review that was held at the end of that year's fall semester and with the student's help we convened a day-long conference composed of those concerned with Yale's uneasy, and not always positive, symbiotic relationship with the city of New Haven. The Saturday morning session included round-table discussions between those constituents at Yale who already had a provable record of working with problematic social issues. Representatives from Yale's Law School, School of Nursing, School of Forestry and Dwight Hall (founded in 1886, Dwight Hall is involved in social-justice issues) joined with representatives of the School of Architecture in order to discuss how they might constructively work together to achieve more wide-ranging solutions to town-and-gown problems than each of their independent operations had been able to achieve autonomously before that time.

Yale–New Haven symposia, New Haven, Connecticut, 1993

During that Saturday afternoon's symposia Yale's president Rick Levin, the university's provost Linda Lorimer, architecture school dean Fred Koetter, key senior faculty members, students and an interested public together with New Haven's mayor John De Stefano Jr. and selected town council members were brought together in the atrium of Paul Rudolph's Art and Architecture Building (in 2008 the revitalized A & A Building was renamed Paul Rudolph Hall) in order to deliberate publicly on the subject of ethics in the context of the many-level relationship between the city and the university. It was an ambitious first of its kind, and the resulting dialogue between those present raised subjects of common concern.

After 15 years, it's still unclear to me if much of substance was articulated at that initial plenary session. However, the very fact that several parties responsible for Yale's and New Haven's joint and several destinies came together signaled that perhaps subsequent dialogs might be beneficial to both institutions.

*　　　*　　　*

Back in Chicago in the autumn of 1994 we convened our first one-year, post-professional multidisciplinary program. It consisted of eight students from diverse educational and vocational backgrounds, in what Bob Somol referred to at the time as existing in a "post-disciplinary" setting. Thus, our 15-year adventure into alternative design education began.

After four years of inhabiting our basement space rent-free, we learned that our landlord had shifted gears and now wanted to charge us rent. In response to that fiscal challenge, one of our trustees, Howard Conant, then generously donated a conveniently located trapezoidal parcel of land that was otherwise unusable for the parking lot it had originally been intended for. Another of our trustees, Judy Neisser, who in addition to her generosity to the school over the life of Archeworks subsequently underwrote the cost of construction of a reasonably priced shed whose shape was governed by the need to take full advantage of its area by filling the entire site. We then built up our full-time staff. Our intern population over the next 15 years slowly increased to an average of between 15 and 20 students and our teaching staff increased to eight.

I can't say that this decade and a half has been without missteps, nor has every project undertaken produced exemplary results, but these years have been filled with an overarching optimism and oomph that we were on the track of something that we felt needed to be done. Neither am I prepared to stipulate that Archeworks is a permanent fixture in the firmament of design education. But it is clear from encouraging individual, collective, institutional and governmental responses that something of value transpired. We were inspired to stretch to the limit transformative ideas about alternative design education because we understood that our graduates would be a conduit to the world of architecture and design practice and as such could introduce alternative mechanisms by which design and architecture could come about.

*   *   *

Throughout the history of architecture, design and art education, certain idiosyncratic institutions have arisen provisionally as a response to societal needs, and have lasted the time necessary to prove their point, and their value, such as Walter Gropius's Bauhaus, Andy Warhol's Factory, Peter Eisenman's Institute for Architecture and Urban Studies and Sam Mockbee's Rural Studio among others similarly constituted. As long as society supports ethically driven, albeit alternative educational institutions devoted to design, safe havens like Archeworks will continue to materialize, necessarily filling a void. Archeworks was as much about a new educational idea generated about the "seam" between the academy and practice as it was about a place where objects were generated and designed for use by those most in need of them.

Ours is a time when programs like Archeworks that unite disparate disciplines seem significant if only by the absence of others. Yet, it strikes me that architects and designers are obligated innately to cooperate in giving something back to society. It would be awful to think that ethically optimistic disciplines that are so bold as to create something out of nothing are undertaken principally to pluck commissions like ripe apples from a tree. Jobs that come about as a result of marketing one's services and branding another's, where advertising is a signification of success, are encouraged by a free-based capitalist society.

Archeworks was conceived of as a notation of a flawed epoch, such that an alternative design school such as ours seemed sensible to us as a possible curative. What has been accomplished there constitutes a record of a time and a place that could somehow absorb an idea like Archeworks as a modest proposal for a certain kind of alternative design education to conceive of the best possible designs for those most in need of it, and then to distribute those very same products to the constituency with whom we had worked.

And so, after virtually a decade and a half in the trenches, whatever else transpires in the future, it is clear to me that an ethical construct underpinned the concept of Archeworks from its beginning to date. I am heartened that architects and designers from all of our related, fundamentally optimistic design fields came together to collectively support and indeed to address the problems of the day. If, in some fashion, Archeworks can be understood as an ethical exemplar, it is because architects and designers responded to the needs of society, personally as well as professionally, and that in effect all by itself made the entire exploration worthwhile.

*   *   *

In 2002 I was invited to be on an architectural jury to select an architect to design a heritage center in Kiev in the Ukraine. The competition was sponsored by the American Jewish Joint Distribution Committee (JDC), a well-known international organization that came into existence for the purpose of supporting European Jewry after the Second World War. Because of my work on the Holocaust museum in Skokie, I went, in part, because I had never seen Babi Yar, the infamous ravine in Kiev into which the bodies of murdered Jews were thrown by Ukrainian paramilitary groups who worked under Nazi direction as the Wehrmacht traveled eastward to engage the Russians. Subsequent to the German's arrival in Kiev, they recruited Ukrainians who were only too willing to manage the carnage. Under this supervision, on September 29th and 30th, 1941, Sonderkommando 4a of the Einsatzgruppe C under the direction of Paul Blobel put to death 33,771 Jewish natives of Kiev. Every Friday thereafter, Jews were rounded up, killed and dumped into that iniquitous ravine until their numbers eventually approximated 100,000.

While I was there for other reasons, as an outsider I stood silently at the edge of Babi Yar one morning watching Kiev residents go about their business as they walked to work apparently impervious to what had transpired at the hands of their predecessors six decades earlier. Returning to my work on the jury, I observed both Ukrainian and Israeli architects making their presentations before the jury selection was made, but I was never free of the image that lingered in my mind's eye of the atrocities committed more than half a century earlier.

*   *   *

In that same spirit that memory is sometimes superimposed upon one's consciousness to unexpected results, Margaret and I traveled to Hungary in 2004 at the behest of a Jesuit priest-cum-architect urban designer who was teaching there as a visiting critic. We had celebrated his work earlier at an Archeworks graduation, when we conferred on him an honorary diploma for his ethically driven approach to urban design issues. In turn, he invited me to be on his jury at the architecture school at the Budapest University of Technology and Economics. He also informed me that his work in Hungary regularly took him to his Jesuit mother's house in Miskolc in eastern Hungary near the Ukrainian border. Known as Transylvania, all four of my grandparents had emigrated from this region, and indeed two of them from this city, more than a century earlier.

At dinner one night in a romantic restaurant set picturesquely among buildings on one of the hills on the Buda side of the Danube, Margaret and I spent a delightful evening consuming considerable amounts of (sweet) Tokaj wine while dreamily listening to a string quartet's syrupy rendition of Czardas. Filled with the gemütlichkeit of the moment, my knee-jerk reaction to this romantic context was to suggest to Margaret that we celebrate my impending 75th birthday in Budapest importing family to explore their roots and friends to help us to understand them.

Ink-on-paper sketch of heritage-center jury, Kiev, Ukraine, 2002

The next morning in the cold light of day while we stood at Buda Castle atop Buda looking down on the Chain Bridge that spans the Danube, I mulled over the atrocities that the Arrow Cross had committed there in the winter of 1943–44. Coming down the mountain and standing on the bank of the river adjacent to the Chain Bridge, I further reflected about the manner in which large numbers of Hungarian Jews had been killed. Lined up at the Danube's edge, these poor souls were handcuffed one to the other. To save ammunition only the lead person in the row was shot, thereby dragging the rest into the water thus drowning the entire line.

Ink-on-paper sketch of Buda Castle atop Budapest, Hungary, 2004

With all due deference to the homeland from whence my grandparents migrated 110 years earlier, we properly celebrated my 75th birthday in Chicago, the place of my birth.

*   *   *

Near the end of the 20th century, the strain that accompanied the conflictual aspects of my personal and professional decision-making combined with the aggregated pressures of practice finally caught up with me—it was time to see a shrink!

As opposed to the ancient WASP psychiatrist who Margaret saw decades earlier as she struggled with the distressing decision of whether to marry me (incidentally who rarely spoke and sometimes catnapped during her sessions with him), my luck was to get an old Jewish Freudian who almost never stopped talking. I kept reminding him that he was schmoozing on my dime, but to no avail. Lest this be perceived as a self-serving rant against psychoanalysis, the old guy did teach me to remove "attitude" while still

245

presenting my opinions with forceful clarity.

Finally, after five years to the day since I began those sessions and significantly on my birthday, when my hour came around, I told him that I was going to use the chair, not the couch in his office. I think he realized that something momentous was about to come down. I reminded him that five years had passed since we began therapy, the precise time it took to get an architectural degree straight out of high school, and that I had decided that I was going to graduate that very day. After we finished, I anxiously awaited the elevator to descend to the lobby taking me to my freedom. I have rarely felt so liberated.

*   *   *

In order to promote the architect rather than his or her constituents, the architectural discipline is drowning in a roiling corporate sea of marketing and branding. I always thought that architecture had long since been enjoined by society to ennoble culture rather than to diminish it by privileging the commercial aspects of design. After all, responsible conduct is essential to those who profess to design "for the good of the many" which happens to be the language employed by the state even as it challenges licensing architects to practice their profession. I always interpreted the mandate of the state as being ethically derived, but that's my own personal interpretation.

My abhorrence to branding or for that matter to marketing, besides genuinely believing that these pursuits diminish the discipline, is based on my personal fear of being compartmentalized, or at some level to even being defined or, even worse, being understood thereby becoming passé. I have always resisted being assigned to a category thinking that architecture is surely more adventuresome than any act of containment. While remaining an unknown quantity has its downside, it is by definition challenging to begin anew each time one emerges at the starting gate. In any case, I simply don't have the patience to pursue a course that necessarily requires the kind of focus that has a personal payoff resulting from product recognition, that is, signature work.

Being an unknown quantity is both a benefit and a burden: a benefit, because it forces one's imagination to the surface, thereby not relying on one's own antecedents; a burden, because of a fear of the unknown by those who might otherwise engage you. Obviously, I believe that the

Ink-on-paper sketch of corporate architects "Have T-square will travel" cartoon, 2005

benefits outweigh the burdens, otherwise I wouldn't have continuously challenged myself in that manner throughout my professional life.

*   *   *

One way that architects can "give back to society" is to go out of their way to work with clients who need their services rather than to ply their trade for those who hire them merely because they can afford them. Often, those who genuinely require the services of an architect don't have that option because they can't afford to hire one. It would be refreshing indeed for marketing directors to seek projects that are need-based.

In the realm of research, it would be useful to the greater collective if architects devoted a portion of their time to addressing the problematic issues of the day such as affordable housing, sustainability, rapid response to natural disasters, et al. I am not, however, referring to the social engineering that helped in defining design in the late 1960s, but rather one that while it is within the context of ethical behavior, is grounded in aesthetics.

*   *   *

With respect to practice in the context of need rather than greed, I will focus here on one recent project that is representative of a particular typology within the general category of institutional work that is in need of proactive involvement by architects for whom design informs their work. The overarching subject is "homelessness," and as absurd as it might seem, in a society like ours that is otherwise presumably sated with a surfeit of unprecedented riches, homelessness remains a problem of some magnitude.

The Pacific Garden Mission (PGM) is an evangelically administered faith-based facility that provides spiritual guidance and essential services for homeless men, women and children without regard to the condition of those who need such support. PGM has operated continuously in Chicago for more than one 130 years and for the past 90 years the male portion of their constituency has been in the same location just south of the Chicago Loop in an area formerly known pejoratively as skid row.

Whitney Griswold, Yale's president during my time in New Haven, once said while in my presence that "no group ever created art", and since that statement was made while he was in the presence of a number of architectural students, I understood him to mean

Ink-on-paper sketch of architects designing "for the good of the many" cartoon, 1994

architecture and the group practice to which he referred was Skidmore, Owings & Merrill. While President Griswold's statement may have been true at a theoretical level for a certain, generally smaller category of buildings, I would nonetheless argue that very few individual architects since the Renaissance have ever built anything of consequence by themselves.

An individual artist might both envisage and execute a painting of even monumental size, but actualizing architectural ideas, no matter how modest they may be, is characteristically the work of many. One of the numerous challenges central to this discipline is both working alone and in concert with others. There is, however, nothing inherently problematic about engaging in both activities at the same time. The PGM project is an example of a project that came about through the efforts of both the individual and the collective.

Something else evolved through my work with the PGM project that opens up rather than closes down opportunities. In some way, design can be interpreted as the purview of the busybody. Because I feel that architecture is inclusivist and not simply "architecture for architecture's sake," an inclusivist architect always goes beyond simply representing form that follows, or perhaps even parallels function. The inclusivist feels that he/she is obligated to suggest programmatic modifications that might impact those for whom the structure is designed to serve. In other words, an architect is more than simply the pencil for a client who produces a program and expects nothing more than the fulfillment of that program.

In the case of PGM, certain features that I suggested to that facility's leadership that might improve life for the homeless were understood by those who administered the facility as something of value, which reinforced the religious underpinning of their faith-based operation and could assist their ministry to help their constituents develop feelings of being useful in a world that otherwise dealt with them at arm's length. These suggestions involved providing greenhouses to tutor the mission's "program people" (those itinerants who signed up for periods of time up to one year in length) for future careers in gardening. The mission's greenhouse(s) could conceivably grow produce for their own consumption, generate organic composting for sale, organize a farmer's market to distribute product to the general population in the facility's cloister and perhaps even penetrate the herb market providing for Chicago's many ethnic restaurants.

Another suggestion was to create a major "interior street" that could encourage interaction between PGM's "overnight guests" and those committed to a longer stay at the facility thus encouraging long-forgotten, or never-attained, social skills. The "yellow brick road" as it is called (the street floor is yellow), houses job fairs and serves as a distribution site for valuable take-away information that makes the city more user-friendly than it is currently.

Since the homeless could be construed as street people, and because a major metropolitan center's streets are anything but user-friendly, it seemed important to create one that would be just that. Street lamps, street signs, street furniture and waste receptacles were employed to reinforce the street image. A security node

Pacific Garden Mission cloister, Chicago, Illinois, 2008

provided a watchful eye on the street and interaction has run rampant on the "yellow brick road."

We designed a landscaped outdoor cloister and broad terraces on upper floors in order to protect homeless persons from the unsympathetic general public as well as to secure quality outdoor space for their perambulations. The cloister provided light to the "yellow brick road," a center for baptisms, landscaped seating and a vista from the small chapel that faces upon it.

We also provided many more counseling rooms, seminar rooms, classrooms and lecture halls than were available in their earlier building. The extra rooms were required to allow PGM to bring as many homeless persons as possible back into the mainstream of society through a broad assortment of educational programs offered.

I felt compelled to suggest these additional benefits because of the often unthinking ways in which civilian populations dismiss, often out of hand, aggressively panhandling homeless persons. Observing the humane ways that the mission staff treated its "overnight guests," I thought it was essential to expand upon that attitude and, in any case, my enthusiasm for the project related to my belief in the value of working in an area of need.

The quantity of outdoor space that came about due to the building's design caused a spike in the building's perimeter, which when combined with the very wide indoor street added to the cost of a structure that might have been more stringently addressed. Nonetheless, the faith-based administrators understood the importance of these value-added elements and without hesitation approved them.

*   *   *

But first, I digress here to make clear how my connection with the PGM came about in the first place.

Though PGM had been located in the same area for 90 years, their immediate neighbor to the north was a well-regarded Chicago public high school, whose constrained site had limited the number and kinds of programs that the school administrators could make available to their students. For years, the Chicago Board of Education generally and that high school specifically coveted the land upon which PGM was situated in hopes of expanding their already well-thought-of programs and services to better serve children of high-school age in Chicago. However, the City could not exercise "eminent domain" over a religious institution.

Frustrated, the Chicago Board of Education assigned a large architectural firm to work with the City's Department of Planning to assist PGM in finding a new site. However, every time the team came upon a site that seemed suitable for PGM's purposes, residential neighbors raised the ugly specter of Nimby (not in my backyard). PGM (incorrectly, as it turned out) feared that only the course of eminent domain remained open to the City of Chicago as a viable, albeit radical option in order to make the mission's land available to its neighbor.

At that point, the mission asked its site-selection architect to represent their position in court should legal action become necessary to resist being displaced. When the architect assigned to them demurred, I was asked to support PGM's position should the issue be brought to trial. My willingness to do so eventually led to a more permanent relationship between us. Our initial efforts together resulted in finding an appropriate site in a ward not immediately proximate to residential development. The second outcome was that in 2004 I was asked to become the architect for the new project.

Embarking on the project, it soon became obvious to PGM's trustees, many of whom were evangelical ministers that my passion for the project carried me far beyond the place where architects normally go. The prime example of the trustees' empathy towards my emotional attachment to the project occurred at a meeting in our office conference room. After I presented several unexpected ideas to the board that were directed at the betterment of their overnight guests, the PGM board chairman interrupted my presentation by asking me if they might segue briefly into a prayer session so that "they could thank God for me!"

*   *   *

If environmental accountability through sustainable architecture was one apparent rationale for the PGM project, another was providing as many options as possible for those who had been shunted aside by society. One reason that "providing architecture for those most in need of it" is compelling to architects is its subjectivity. The designer identifies with a particular constituency, perhaps fearing that, "there, but for the grace of God, go I."

The PGM project occurred at a time later in my life when, as many of us do, I felt the need to revisit an earlier language that had informed my work in the past. That language, the Chicago School's bare-bones exposition of a building's structure seemed appropriate in the context of a project where sparsity was not only called for but was also a reasonable representation of a pared down budget. Rather than indulging in a style that might have suggested a "homey" quality replete with domestic cues such as pediments, mullioned domestic appearing windows, residential scaling, et al, I felt it was essential to come up with a no-nonsense, uncomplicated architecture that eschewed the maudlin in favor of representing fairly the grit of the city. It is important to understand that PGM is not a home for homeless persons. To have provided architectural cues that represent "home" would have been dissimulative.

The exposition on the façade of the building of concrete slabs supported by concrete columns that represented an approximate 20-by-20-foot structural bay system had the additional benefit of reducing the amount of exterior materials that would have been otherwise necessary to enclose the building. Given the cloistered central courtyard that added additional perimeter to a more conventionally defined rectangular parallelepiped, any system that might reduce costs on the periphery seemed appropriate.

*   *   *

John Hejduk put it admirably when he once suggested that "architecture is a passionate pursuit." The idea that there is a Zen-like inference of being at one with one's work is implicit in Hejduk's notion. In that spirit, an architect might well think that distancing one's self from architectural production is not the most humane approach to designing and building for those in need. If, for some inexplicable reason, one is compelled to establish distance between subject and object, it might be more logical to be detached from an intellectual position, or from an esoteric architecture rather than being detached from one's own emotive state.

*   *   *

In the spirit of John Hejduk's strong belief in "architecture as a passionate pursuit," there are few architects practicing today that better exemplify designing and building from an emotional point of view than Rome's own Massimiliano Fuksas. Through form and color, but mostly through attitude, Fuksas has added an expressive element to architecture seldom seen in the 20th century, if one excludes Frank Lloyd Wright and John Hejduk himself.

The title of Alberto Pérez-Gómez' book Built upon Love: Architectural Longing after Ethics and Aesthetics, might well sum up one's fascination with passion in the context of a humane architecture. However, this is not a plea that is opposed towards ideation. I just think that designing has the potential of being at once intellectually informed and, then again, something more instinctive than that.

When UIC's architectural historian and theorist Annie Pedret read her version of "Ethics" for *The Archeworks Papers* (published as vol. I, no. 4 in 2006–07), I was reminded of both John Hejduk's and Alberto Pérez-Gómez' introduction of the topic of love into architectural theory and practice. Hejduk's understanding of love as seen in his drawings, his buildings and his poetry was emotionally driven, much in the way that the contemporary Italian architect Massimiliano Fuksas' work flows from the same wellspring of feeling. Unfortunately, Pérez-Gómez' book, as do his earlier writings, reduces the subject matter to formula.

While Pedret's approach to the subject of ethics is sequential—first "being," then "understanding" and finally "action"—I believe that a more proactive approach is required in order to come to grips with all three stages. Pedret argues that, "only by voiding the mind of predilections can one induce a state by which the mind can be at one with being, understanding and action." On the other hand, I believe that only by adopting a proactive posture leading towards habituation can ethical behavior move vigorously in the direction of "making a difference."

The legacy of what one leaves behind

I have never understood how a noble discipline such as architecture has produced such a plethora of practitioners ostensibly devoted to their own well-being (commerce, professionalism, even art), when it is otherwise so clear to me that their responsibilities to the society that nourished them transcends self-satisfaction. After all, at the end of the day, what is it that one leaves behind? For an architect, "good works" includes attitudes embellished by drawings, buildings, texts as well as former students, that is, the byproduct of mentoring and/or teaching. Mostly, one's legacy is informed by how others have perceived the way in which one has had an impact upon one's world. In a word, "reputation."

The field of architecture is one of those disciplines that expose those who practice it to a level of scrutiny so transparent that it is difficult to conceal one's motivations. While not fundamentally problematic, for those whose interest is directed toward the self—such as personal gain in one form or another represented by seeking profit, notoriety, security, even aesthetic adventuresomeness—such motives soon become apparent. While not a priori laudatory, for those whose career trajectory is directed towards different ends that are focused outside the self such as building well, giving back to society, behaving ethically, providing services for those most in need of them, those features soon become obvious as well.

The option to select one or another direction is both to the point and discrete, and whichever trajectory is chosen, one can rest assured that the choice one makes not only defines character, but makes clear to all the motivation by which decisions are made. While others in a capitalist society are rarely confronted by the incentive underlying their choices, those disciplines representing culture are held to a higher standard.

One of the constituent features defining character is embedded in an architect's ability to be courageous when it counts. The ability to "stick to one's guns" is sometimes tested while in architecture school through the mechanism of the jury system, by which one often rises or falls based on the perception about the strength of one's convictions. The following anecdote addresses the subject of courage.

In the mid-1980s, a former student of mine came to see me in my office. He didn't stipulate what he wanted to talk about, but teachers and students, much like parents and children, enjoy a kind of extra-sensory perception vis-à-vis their interdependence. Since I knew that he had been employed for some time as a draftsman, and that he was increasingly unhappy in that position, I had a pretty good idea what he wanted to discuss.

Without any preliminaries, he asked, "What are my chances?" Of course, he was asking whether or not he would succeed in independent practice, should he decide to go out on his own. I responded by saying, "If you have a modicum of talent, if you are willing to work endless hours, if you are lucky, and if you are willing to sacrifice whatever is needed, then perhaps 10 to 15 percent." He thanked me and without further ado left our meeting, and immediately opened up his own architectural practice. Thirty years later he is still in practice. But it was his courage to forge ahead into unforeseen territory in spite of the odds that made me proud of him.

*   *   *

I would argue that there are other vehicles better suited to reflecting an orderly existence than architecture. Unlike Mies van der Rohe, I feel that architecture needs to challenge a given epoch, rather than to simply synthesize it by "reflecting it" (re: Mies's 1924 well-known quote, "Architecture is the will of an epoch translated into space"), particularly when the society itself is enmeshed in conflict. I would further submit that every epoch thus far has had its share of conflict. This is not to suggest that expressing conflict for its own sake needs to inform architectural design any more than synthesis is itself a predominantly appropriate architectural expression. I am, however, suggesting that society in one way or another has in some measure always been conflicted and that architects need not shy away from expressing conflict in their work, wherever and whenever they feel it appropriate.

At the same time, architects are also obligated to "show a better way." It is imperative that architects attempt to heal societal rifts through forms that they help to bring about. Because architecture is, more often than not, the result of a synthesizing process that includes many subsets such as function, use, construction, structure, and site that tend to be subsumed into an overarching solution that reinforces the reputation of architecture as "the mother of the arts," there is always the temptation to embrace the category of work that is framed by synthesis as the one and only way in which to build. The synthesized solution does seem to represent showing a better way, whereas the representation of conflict is best achieved by expressing the disparate parts of a problem without subsuming them into a synthesized whole.

The dilemma of seemingly ostensible opposites appears once again, raising its head as if to perplex the practitioner. Yet, contradistinctive ways of expressing society through the vehicle of architecture always challenges the practitioner to seek his/her own way so as to bring their belief systems to come to consciousness in sharp contrast.

These kinds of analyses strike me as ultimately favoring conflict over and against synthesis as a mechanism of architectural expression, if only to better represent the unresolved struggles that humankind regularly faces without providing answers that, in any case, can best be ascertained without reducing those answers to architectural formularization.

Classicism versus romanticism

From a formalistic point of view, architectural history tends to be denoted by a pendulum that swings back and forth between classicism and romanticism. Over time, classicism became the dominant force since it is the more dependable of the two. I might add that it is the more inclined to formularization, reiteration, et al, therefore occupying more time in the collective minds of the architects of any given age than its opposite number. Every so often, however, when classicism's constancy becomes excruciatingly tedious, the delightfully unpredictable head of romanticism rears up to, albeit provisionally, displace it.

In any case, it is clear to me that architects never solve problems through the medium of their work, they merely express problems by helping to bring about structures that extend programs into being. If expressing "the will of an epoch," as Mies van der Rohe pointedly referred to it, is largely technologically driven, that's one thing. Of course, the forms that Mies used to express the epochal will were rectangular parallelepipeds, whose flexibility was beyond dispute, therefore encouraging their usage in differentiated typologies for an inordinate period of time; which is precisely what classicism has been about throughout history.

If however, that same "will" is more comprehensively evident in societal struggles, the resultant architectural forms that are utilized for representation are more complex and significantly less paradigmatic; which is precisely what romanticism is about. Needless to say, since romanticism is capricious and individualistic and thus difficult to emulate, it is noticeably less long-lived than its classical counterpart, therefore suggesting perhaps that classicism is more meaningful over the long term than romanticism.

Without question, it is difficult to sustain a career that is wholly romantically driven. Such a career focus requires a mentality not often found in creative persons who have given over their life to a discipline that is, more often than not, born into usefulness and thus informed by pragmatism, such as architecture. But without sensitive souls such as Borromini, Frank Lloyd Wright, John Hejduk, Frank Gehry and Massimiliano Fuksas, architecture would lose an essential dimension that represents the emotional component of life seen through the lens of romanticism represented in form and material.

Another way of denoting the difference between romanticism and classicism is to describe the difference between Daniel Burnham's 1909 plan of Chicago and the Jeffersonian grid superimposed on the flat plane of the prairie abutting Lake Michigan. Burnham's plan is European in origin insofar as it is hierarchical in nature. The plaza that gives way to the cathedral and the square that fronts the town hall are romantic picturesque notions that embellish the importance of the church and state, while the grid within which populations dwell is something of an abstraction that is at once democratic and alienating.

In summary, classicism tends to be normative and expected. Romanticism represents the one-off, the "black swan." Nassim Taleb defines the unpredictability innate as the "black swan," which in architectural terms can be defined as romanticism.

*   *   *

I'm not entirely persuaded that architecture schools equip students with the philosophic tools to ethically make what are inexorably complex decisions that convincingly contend with issues that are often mutually exclusive. Thus, it can be argued that architectural education itself can be called into question.

Is the primary purpose of architectural curricula focused on producing graduates useful to future employers? Vast amounts of time seem to be ever more required to equip architecture students to be at ease with incessantly expanding technology, leading one, in our own epoch, to the conclusion that the purpose of architectural education is to produce graduates who are increasingly user-friendly to computers and other technological processes. Which is to say that other perhaps more important issues are left by the wayside in an educational system whose duration is, in any case, woefully limited. It seems to me that the purpose of architecture school is not only to give students the tools to be useful to employers in the field, but to equip students with the comprehensive abilities to make decisions that, more often than not, are ethically complex in nature. Insofar as ethics is arguably more important educationally than familiarity with the latest computer software, it strikes me that ethics requires greater emphasis educationally than it is currently given.

But lest this become an antagonistic tirade directed at the ethical shortcomings of architectural education, one must first be comfortable with ethics in the framework of one's own self. And that is something that precedes the education of an architect. Socrates once posited the idea that ethics is a function of "habituation" more than anything else. For that matter, one can argue that "value" is an intrinsic component of architecture that, in order to be understood, one needs to understand the innate nature of the concept of "good," both from a relative, as well as an absolute, point of view. G.E. Moore's early-20th-century treatise Principia Ethica is an essential primer describing the difference between relative and absolute value.

In all events, referring to Socrates' belief in the value of habituation, since architecture requires practice and if ethics is intrinsic to architecture, it follows that architects are innately habitual in their work. They only need to be mindful about the importance of habituation to raise the bar in their work.

Inconclusive musings

Something of consequence transpired during the four years that I was preparing this text, which illuminated an important lesson. Throughout my life as an architect, I have placed great emphasis on the juxtaposition between teaching and doing. Like so many mothers, mine also said that, "those who can do and those who can't teach." In this text, I stipulate that teaching is fundamentally generous, that is, the empowerment of a symbiotic other, while doing is, at least in some measure, about giving to one's self.

In the course of organizing my thoughts on this dialectical duo, I came to understand that ever since

I had shifted my practice in the direction of projects where "poignancy was built into the program," there was in fact a not-inconsequential component of giving that had found its way into my doing. Before my career trajectory had spun off in that direction, projects came my way without rhyme or reason. Since I never marketed my work, my practice, much like that of other practitioners, was subject to the vicissitudes of the times and the idiosyncratic behavior of those clients who walked in the door. God bless them, each and every one. With the exception of those clients whose mishegoss was beyond the pale, I accepted virtually all those who came calling.

*   *   *

With the coming of the new millenium, I became aware of certain concomitant factors in practice that suggested a dialectical state between interiority and exteriority that I was never completely mindful of previously. Between 1968 and 2008 a dozen projects had come about in my practice that embodied, in one way or another, this duality. Many, but not all of them were theologically based projects, but all of them related in some ways to Cistercian monastic precedent. The number 12 here is thought to be related to the 12 tribes enumerated in the Old Testament or the 12 apostles enumerated in the New Testament, depending which Bible influences your understanding of the significance of the number 12.

Three of these projects—St. John's at the University of Illinois, The Baha'i Archives in Wilmette, Illinois, and the IBA villa in Berlin—are cleaved, the resulting spaces of which at least imply interiority even if they are not entirely cloistered spaces. Three others portray "inaccessibility" through an orientational shift away from a Cartesian precedent and towards the West Wall of the temple in Jerusalem: the DAM Niche in Frankfurt, Germany; the Or Shalom unbuilt scheme in Metawa, Illinois; and the Illinois Holocaust Museum and Education Center in Skokie. The final six projects articulate completely contained cloisters well within the Cistercian tradition: the Momochi housing in Fukuoka, Japan; the Educare preschool center and the Children's Advocacy Center both in Chicago, Illinois; the unbuilt interfaith proposal in San Francisco; and the Pacific Garden Mission and the U.C.A.N. Campus, both in Chicago, Illinois.

All 12 of these projects can be understood as representative of the difference between exteriority and interiority, or in other words the public face versus the private face, or practice (public) versus teaching (private). I refer to these 12 projects because they depict a change in my understanding of architecture as I attempt to invest the architectural discipline, at least the very small part that my practice plays in it, with at least a part of the energy that I had formerly put into teaching.

Stemming from my knowledge of Cistercian monastic precedent based on my study of such cathedrals as the ones at Cluny, Cologne, Durham and Wells, that is, the presence of an exterior face to the world versus the cloistered world with its privately spiritual axis mundi, which produces a discourse not dissimilar from the one that I understood as architectural practice versus teaching.

Simplistically, buildings present a face to the public that only in some ways can signify what is behind that façade. Sometimes that public face is an accurate representation of what lies beyond, sometimes it is a neutral representation. Being contextually responsive is a rationale often given by those so concerned. Sometimes that face is actually a mask concealing as much as it reveals. When the public face is deliberately something "other" than what it conceals, the private life behind the façade is just that—private.

The cloistered world that is both exterior and interior and yet open to the sky is, at least in a theological sense, when the use of the building addresses an ecclesiastical purpose, an axis mundi, an aspiration to be at one with a heavenly divine being. In any case, such a cloistered space is spiritually inclined, at least in the way that I understand the inner self as expressed in a poetic sense, which is to say an aura or an ineffable component that could add an important dimension to one's method of working.

Most importantly, I discovered the importance of an inaccessible void as a perpetually unfulfilled desire—loosely considered, a space that might be perceived as unheimlich. No matter how satisfying an architectural space might be, once it is inhabited the space necessarily loses some of the aura that one might have assigned to it when it is simplistically situated within one's imagination. Once such a space is actualized, however, it loses some of the magic, or mystery, before one actually inhabits it. It may be that the mythos of architecture curiously is diminished by the simple fact of inhabitation.

On the other hand, the cloister that is marked by an inaccessible center—or more, the inaccessible void—retains its mystery precisely because of the impossibility of thoroughly "knowing" it. The dozen projects that I conceived with a cloistered center, many of which are uninhabitable, gave me an understanding of the importance of such an approach to design; which is to say that holding back from knowing everything may have value not previously understood by the practitioners of the discipline.

*   *   *

Once I made the decision to work primarily with clients with real, not imagined, need I soon realized that the time that I had assigned to teaching, I was able to redirect to practice, and that fragmentation, that conflict, was no longer as essential as I once thought it was.

Furthermore, given the preference between teaching and practice, at the end of the day I am, more than anything else, a practicing architect. I like nothing more than attending construction job-site meetings. This tendency goes back to my beginnings in this field, remembering that as a teenager I was employed in George Fred Keck's office where the maestro instructed me to concentrate more on the actualizing side of this profession rather than the more design oriented side.

In reality, however, I may be a better teacher than I am a practitioner, but perhaps because of my own perversity I was able to engage in a self-seducing state to clarify some things in my mind that in turn led me to resign from Archeworks after 15 years of directing that institution. Thus, for the first time since I returned to Chicago from graduate school, I am able to devote my entire energy to practice.

Paradoxically, my resignation from Archeworks coincided with my being awarded the Topaz Medallion for a lifetime of teaching architecture. This prized medal is given jointly from the Associated Collegiate Schools of Architecture and The American Institute of Architects and is conferred on one person annually. In my view, however, I received that honor under false pretenses. Though my age may mistakenly qualify me as an éminence grise as I stipulated earlier, I am neither Socratic nor an intellectual, thus I am not a typical academic by any stretch of the imagination. What I am is a practicing architect who happened to teach architectural students for the four-and-a-half decades that I practiced the same discipline that I taught.

*   *   *

My personality is such that, when I am involved in the early creative years of any initiative, I am far more animated by the process than in later years when the management of an established bureaucracy consumes the energy that I felt at its initiation, causing my interest to flag. When we began Archeworks, the school represented a revolutionary idea that created great excitement among those of us involved in its development, and later in its refinement. A decade and a half after its inception, our adventure into alternative design education seems to have made whatever mark that it might make. Other institutions have begun, in some ways, to emulate it. In any case, it is not in my aesthetic DNA to "clip coupons" on situations requiring institutionalization (read bureaucratization).

I am hardly the first person to be less than enthralled by certain kinds of repetitiousness irrespective of my touted ability to happily engage in mundane tasks that do not require a great deal of thought. It is said that when the Italian sculptor Alberto Giacometti was on his deathbed, he reportedly admitted that he was sorry he had ever made "them;" he was referring of course to his well known distended bronze figures. He seemed to be suggesting that he regretted his lifelong commitment to self-similar sculptural forms that he felt unnecessarily limited his ability to express himself more fully than he did. Of course, one will never know where and to what extent Giacometti might have taken the implied independence for which he allegedly so yearned.

No longer being continuously involved in architectural education, I am forced to make sure that my practice from an ethical point of view remains true to its set course of working with those whose need of architecture lies within the context of social cause. To fulfill my aspiration to embed educational conceits into the practicum necessary for their accomplishment would be sufficiently rewarding.

*   *   *

One final not-so-startling observation, having built far afield from my home base on any number of occasions in locations from Bangladesh and Japan at the farthest, through much of Europe, Canada, Puerto Rico and all across the lower 48, I have no great passion to further indulge my ego by building in other exotic locales. To build far from the source where designs are conceptualized leads to the actualization of diagrams. Such a project cannot benefit from rigorously following through on all the myriad of details that embellish the original concept such that it is transfigured as a complete work of architecture. While I realize that mine is an outdated notion about building, it represents many decades similarly inclined.

The reason that I have reached this conclusion at this late date is the normalcy (call it convenience) that we all, each in our own way, seek as we move inescapably towards our own endgame, imploding as we inevitably will so as to find comfort levels suitable to our diminishing capaciousness. In other words, more often than not it is useless to insist on the same method of operation that served our purposes at earlier, more vigorous stages of our life. In this later stage it is more important to find more useful methods by which we can actualize work.

*   *   *

Thus concludes this story of hopping from architectural trench to architectural trench that, while not extraordinary in most ways, was not always conducted under ordinary circumstances. Is there a moral connected with these episodic reminiscences? Not really, with the possible exception of endeavoring to stay awake and aware while life presents its usual complex choices. Like so many other life stories, these too are filled with errors of judgment or roads not taken.

Epilogue

As a footnote to eight decades of trying to come to grips with both teaching and practicing architecture, in spring 2009 I underwent the first of six surgical procedures for heart-related issues. Three months in hospital gave me pause to reflect about the difference between absolute and relative values; the very same values that I focus on in the text. My health hiatus only confirmed my beliefs.

Stanley Tigerman, 2010

The author at work, Chicago, Illinois, 2005

Acknowledgements

Both practicing architects and architectural educators cannot but acknowledge the vast number of employees, associates, partners—even clients—who together with faculty colleagues, researchers and students not only helped to flesh out one's career, but who also participated—often equally—in the evolution of ideas that otherwise might be mistakenly credited solely to a single author.

I am no exception. While I assume the responsibility for how I may have erred in the compilation of this text, I cannot take exclusive credit for what might be of merit herein.

My apologies go to those who I may have inadvertently omitted from this text, my thanks to those who sometimes inadvertently helped in charting my perpetually changing course and my everlasting gratitude to those colleagues, friends and relatives who stood by me, tolerating my checkered past, my controversial present and, for better and worse, if history repeats itself, my uncertain future.

Mara Wilhelm and Taber Wayne did much of the hard work to bring this to closure, while my publishers did what they could to correlate the whole so it might be perceived as somehow a reasonable reminiscence of being outside.

If at some level JJ and Tracy suffered by my absence, Margaret struggled with my often over-weaning presence—especially during the blue pencil process.

Stanley Tigerman, Chicago, 2010

List of Photo Credits

All photographs appearing in the book are courtesy of Tigerman McCurry Architects except where otherwise noted.

Introduction
P. 14 Photograph courtesy of The Art Institute of Chicago
P. 24 Nick Merrick © Hedrich Blessing

Chapter 1
P. 41 (left) Photograph courtesy of Margaret McCurry

Chapter 2
P. 45 Photograph courtesy of Margaret McCurry
P. 46 Photography courtesy of Markuse Corporation
P. 49 Photo courtesy of Chicago History Museum, Hedrich Blessing Collection
P. 50 Photo courtesy of Chicago History Museum, Hedrich Blessing Collection
P. 56 Photograph courtesy of Rafique Islam
P. 61 Photograph courtesy of Margaret McCurry
P. 61 Photograph courtesy of Margaret McCurry
P. 58 (bottom) Photography by Moholy-Nagy, courtesy of The Art Institute of Chicago
P. 73 Photograph courtesy of the Chicago History Museum, Hedrich Blessing Collection
P. 80 Photograph courtesy of US Navy
P. 89 Photograph courtesy of Margaret McCurry
P. 100 Photograph courtesy of Yale University

P. 100 Photograph courtesy of Yale University
P. 103 Photography by M. Henry, courtesy of author
P. 104 Photograph courtesy of Yale University
P. 115 (top left) Photograph by Balthazar Korab
P. 115 (top right) Photograph by Orlando Cabanban
P. 117 Photograph by Orlando Cabanban
P. 120 Photography by Philip Turner
P. 120 Photography by Philip Turner
P. 121 Photography by David Hirsch
P. 121 Photography by Philip Turner
P. 125 (top right) Photography by Orlando Cabanban
P. 151 Photography by Orlando Cabanban
P. 157 Photography by Orlando Cabanban
P. 158 Photography by Orlando Cabanban
P. 170 (bottom) Photograph courtesy of Margaret McCurry
P. 172 (top left) Photography by Rich Hein
P. 172 (top right) Photography by Bruce Van Inwegen
P. 172 (middle left) Photography by Bruce Van Inwegen
P. 172 (bottom left) Photograph courtesy of ARCHEWORKS
P. 176 Photograph courtesy of Margaret McCurry
P. 177 (top) Photography by Sadin/Karant Photography
P. 177 (bottom) Photograph courtesy of Margaret McCurry
P. 178 (bottom) Photograph courtesy of Margaret McCurry
P. 182 Photograph courtesy of Margaret McCurry

P. 186 (top right) Photograph courtesy of
Margaret McCurry
P. 186 (bottom left) Photography by
Tom Schumacher
P. 189 (middle and bottom) Photograph
courtesy of Tigerman McCurry Architects (gift of
Gulbenkian Foundation)
P. 191 Photography by Peter Aaron, courtesy
of ESTO
P. 192 Photography by Peter Aaron, courtesy
of ESTO

Chapter 3
P. 196 Photography by Robert Lautman
P. 206 (top) Photograph courtesy of Rizzoli New York
P. 207 Photograph by Timothy Hursley
P. 212 Photograph by Orlando Cabanban
P. 215 (top) Photography by Brian Fitz
Photography, courtesy of Bulley and Andrews
P. 215 (bottom) Photography by Bill Zbaren,
courtesy of Zbaren Photography
P. 220 Photography by Bill Zbaren, courtesy of
Zbaren Photography

Chapter 4
P. 226 Photography by Bruce Van Inwegen
P. 227 Bob Harr © Hedrich Blessing
P. 228 (top and bottom) Steve Hall © Hedrich Blessing
P. 241 Photograph courtesy of Yale University
P. 249 Steve Hall © Hedrich Blessing

ORO *editions*

Publishers of Architecture, Art, and Design
Gordon Goff: Publisher

www.oroeditions.com
info@oroeditions.com

Designing Bridges to Burn
Copyright © 2011 by ORO editions
ISBN: 978-1-935935-07-0

Book Design: Usana Shadday
Cover Design: Natalie May
Production Assistance: Gabriel Ely
Project Coordinator: Christy LaFaver

Printed in Canada

ORO editions has made every effort to minimize the overall carbon
footprint of this project. As part of this goal, ORO editions, in association with Global ReLeaf, have arranged to plant two trees for each and
every tree used in the manufacturing of the paper produced for this
book. Global ReLeaf is an international campaign run by American
Forests, the nation's oldest nonprofit conservation organization. Global
ReLeaf is American Forests' education and action program that helps
individuals, organizations, agencies, and corporations improve the local and global environment by planting and caring for trees.

International Distribution
www.oroeditions.com